Bones for Barnum Brown

BONES FOR BARNUM BROWN

Adventures of a Dinosaur Hunter

By Roland T. Bird

Edited by V. Theodore Schreiber

With a Foreword by Edwin H. Colbert
Introduction & Annotations by James O. Farlow

Texas Christian University Press
Fort Worth

Library of Congress Cataloging in Publication Data

Bird, Roland T. (Roland Thaxter), 1899–1978
Bones for Barnum Brown.

Includes index.
1. Bird, Roland T. (Roland Thaxter), 1899–1978
2. Paleontologists—United States—Biography.
I. Schreiber, V. Theodore. II. Title.
QE707.B57A33   1985   560'.9 [B]   84–24047
ISBN 0-87565-007-4
ISBN 0-87565-011-2 (pbk.)

Designed by Whitehead & Whitehead / Austin

The dinosaur on the cover is a *Triceratops*, a horned dinosaur or ceratopsian. Working with related ceratopsians, Barnum Brown and his coworkers in the laboratory at the American Museum made an important discovery by identifying a previously unrecognized bone in the lower jaw.

Foreword

IT WAS MY PRIVILEGE to have known Roland Thaxter Bird during those years when he was right-hand man to Barnum Brown. That was mainly back in the thirties, when Brown was enjoying his well-deserved reputation as a hunter of dinosaurs, and when Bird was his enthusiastic, energetic, and innovative field assistant. They made a great team.

In addition to being a dinosaur hunter, Brown was an indefatigable practitioner of public relations, all for the furthering of ways and means to raise support for his field program. So he was much in the limelight. R. T. (as he was affectionately known to friends and colleagues) thus was much within the long shadow cast by Brown, but I am sure he did not mind. He was a quiet, modest person, and for him the quest for fossils was of paramount importance.

Consequently, R. T. spent many months each year out in the field looking for and collecting the ancient bones he loved so well—part of the time with Brown, part of the time by himself or with some assistants—while Brown was off on his special pursuits. Those days out under the arching sky, so graphically recounted by R. T. in this book, were the adventures for which he lived. For him, work in the museum laboratory was all right but not exactly his first priority. Always he was thinking ahead to the next field season.

But this did not preclude thoughts about the meaning of the fossils he collected. Indeed, R. T. published some evocative articles on the Lower Cretaceous sauropod footprints that he collected near Glen Rose, Texas, and it was my particular pleasure to have collaborated with him on a description of the huge Upper Cretaceous crocodile that he collected from the Texas Big Bend region.

Yet for R. T. such literary excursions perhaps were side trips, not to be compared with his work on the actual fossils. So it was that the last task he performed for the American Museum, the assembling of the Glen Rose footprints in one of the dinosaur halls, was for him an especially happy assignment. Moreover, it was a fortunate assignment for the museum, because only R. T. could have done the work as it needed to be done. Today it is a truly spectacular display, in every way his monument to be admired through the years by millions of people.

As has been said, R. T. was a thoroughly modest man, completely devoted to the fossils on which he was working. At times he was just a bit unworldly. I distinctly remember the occasions when someone from the museum business office would come to the laboratory and on bended knee (figuratively, if not literally) beg R. T. to cash or deposit his latest paycheck so that the museum books could be balanced. R. T. would then shuffle through a drawer where he had the check, or possibly several checks, stashed away, awaiting the day when he could attend to such mundane matters as bookkeeping.

R. T.'s brother, Junius, a noted archaeologist ensconced at the other end of the museum, would now and then drop in for a visit, but not for long would he divert R. T. from his appointed work. Likewise, the other men in the laboratory enjoyed visiting with R. T., who was a friend to all, but the talks were usually brief; R. T. must get on with his work.

The scope of his work, as set forth in the pages that follow, is told by one of the last of the "old time" fossil hunters, one of the men out to collect bones and footprints, unencumbered by problems of taphonomy, sedimentology, the esoteric aspects of stratigraphy, population structures, and other concerns that so often furrow the brows of today's paleontologists. We are all indebted to R. T. for having shared his experiences with us.

We are also indebted to Theodore Schreiber for bringing this manuscript to publication and to James Farlow for his excellent introduction, providing an informative background that should be helpful for those who read this book.

Edwin H. Colbert

 Bones for Barnum Brown

Roland T. Bird

Introduction

ROLAND THAXTER BIRD was born on December 29, 1899, the first of four children of Henry and Harriet Slater Bird of Rye, New York. Henry Bird was a successful businessman, a practical man who insisted that his children learn a trade; thus at one point in his youth R. T. was apprenticed to a plumber. At the same time, however, Henry Bird (or "Pater," as his children called him) had a deep respect for learning. He was himself a distinguished amateur entomologist who did important studies on noctuid moths.

The Bird household was a warm and stimulating environment. As children, R.T. and his sister Alice (the closest of his siblings in age) were each given one of the original Teddy Bears; Harriet made vests for the bears, and R.T. built cars, boats, airships, and even a house with electric lights for them. Alice also had a china doll, and on one occasion she and R.T. decided that the doll needed a baby, so they stuffed tissue up its dress to make it look pregnant—presumably to the consternation of their elders. Pater had purchased first editions of both *Tom Sawyer* and *Huckleberry Finn* and frequently read aloud from them to his children, not infrequently being interrupted by his own uncontrollable laughter. Not surprisingly, R.T. developed a lifelong love for the works of Mark Twain, and in writing his memoir thought of himself as a kind of latter-day, scientific Huck Finn. Given the nature of his childhood environment, it is not surprising that R.T. went on to distinguish himself as a fossil collector, or that his brother Junius became a leader in the field of South American archaeology.

Harriet Bird died of tuberculosis at the age of forty-two, when R.T. was only fifteen years old. The boy's health had never been robust (he frequently suffered from colds), and Henry feared that his son might also contract tuberculosis. Upon the advice of the family doctor R.T. was sent away from the coastal town of Rye to live on an uncle's farm in the Catskills; there he developed an interest in cattle and their husbandry that he retained for the rest of his life.

As a young adult R.T. went to work on a farm in Florida, caring for Jersey cattle and traveling about the country showing them at exhibitions. He invested in real estate during the Florida land rush of the 1920's and, like so many others, lost what little money he had in the economic crash at the end of the decade. After constructing a motorcycle-camper, Bird wandered about the country, visiting all of the states and supporting himself by doing odd jobs; so it was that in November, 1932, at the age of thirty-two, he passed through northern Arizona and discovered the fossilized skull that changed his life.

Bird's career as a dinosaur hunter was short but eventful. He assisted in the collection of some very important specimens, and his discovery of sauropod footprints in the Cretaceous rocks of Texas made him a significant figure in the history of paleontology. To put his career, and that of Barnum Brown, the employer Bird admired so much, into a proper perspective, we need to review the history of the study of dinosaurs.

In 1802 a boy named Pliny Moody found some fossilized footprints in rocks of the Connecticut River Valley near South Hadley, Massachusetts. These tracks were so similar to those of birds that local people referred to them as poultry tracks, or the footprints of Noah's raven. More and more such tracks turned up at an increasing number of localities, and some of these were quite large—as much as eighteen inches long, suggesting "poultry" on a grand scale.

The tracks came to the attention of Edward Hitchcock, a geologist and clergyman who became a professor of chemistry, geology, and theology, and president of Amherst College. Hitchcock zealously collected and described fossil tracks from the Connecticut Valley, the culmination of his work being *Ichnology of New England. A Report on the Sandstone of the Connecticut Valley, Especially Its Fossil Footmarks*, published by the State of Massachusetts in 1858 (a supplement to this tome was published posthumously in 1865).

Hitchcock compared the three-toed tracks in his collection to those of birds and concluded that they had been made by huge flightless birds similar to modern ostriches, rheas, cassowaries, and emus, and to even larger extinct birds like the moas of New Zealand.

It was a perfectly reasonable mistake. We now know that Hitchcock's three-toed tracks were made by bipedal dinosaurs, whose feet were astonishingly bird-like (indeed, most paleontologists now believe that birds evolved from small, bipedal, carnivorous dinosaurs). Hitchcock was aware, and troubled by, seemingly reptilian features of some of his fossil trails. In describing one of them, he wrote: "When I first saw its track, although it had a small fourth toe, I thought it a bird; but when I found soon after that it had left the distinct trace of a tail, that opinion must be abandoned; for the trace could neither be explained by referring it to the dragging of the feet, nor to the large tail feather of a bird. Yet if a biped, its body must have had somewhat the form of a bird, in order to keep it properly balanced. The tail, although evidently rather stout, was not enough so to help prop up the body. And how very strange must have been the appearance of a lizard, or batrachian, with feet and body like those of a bird, yet dragging a veritable tail!"

Strange indeed, but a splendidly prophetic description of a bipedal dinosaur. To his dying day Hitchcock affirmed the avian nature of most of his tracks, although he concluded that the birds that made them might have been somewhat reptilian in character. Hitchcock never knew that his "reptilian birds" were dinosaurs.

And yet he did know that such animals as dinosaurs had once existed. The first dinosaur was scientifically described by William Buckland, a brilliant if eccentric British cleric and scientist; in 1824 Buckland published a paper on the remains of a large carnivorous dinosaur, which he named *Megalosaurus*. The following year saw a description of another dinosaur, *Iguanodon*, this one an herbivore, published by an English physician named Gideon Mantell.

By 1841 several additional kinds of dinosaurs had been found, and Richard Owen, a British comparative anatomist (who founded the British Museum of Natural History), realized that these creatures represented animals rather different from living reptiles—even though complete skeletons were known as yet from none of these dinosaurs. At a meeting of the British Association for the Advancement of Science, Owen proposed recognizing a distinct group of fossil reptiles, which he named "dinosaurs," Greek for "terrible lizards." Owen took pains to show that, despite their formal name, dinosaurs were not lizards but were structurally advanced over living reptiles, in many respects resembling mammals. In part his motivation for this was his staunch anti-evolutionism, evolutionary ideas having been "in the air" at the time; Owen believed that the replacement of the grand dinosaurs of the past by the smaller, less structurally sophisticated reptiles of the present proved that degeneration, rather than evolutionary progress, had typified the history of the reptiles. When in 1854 he supervised the construction of several life-sized models of prehistoric animals for the grounds of the Crystal Palace of London, Owen restored dinosaurs as huge, lumbering, quadrupedal monsters with an erect, pachyderm-like carriage. This was the image of dinosaurs in Hitchcock's mind as he studied the ancient footprints of the Connecticut Valley, and so it is not surprising that he did not attribute the tracks to dinosaurs.

In 1858, the year Hitchcock published his monumental book on fossil footprints, bones of a large fossil reptile were discovered at Haddonfield, New Jersey. These remains were much more complete than any previously discovered, and Joseph Leidy, the American anatomist and zoologist who described them, was impressed by the marked difference in size between the dinosaur's relatively small

Some of the dinosaur tracks described by Hitchcock (1858) and identified by him as tracks of giant birds. The footprint in the lower lefthand corner was interpreted as the track of a birdlike lizard or salamander.

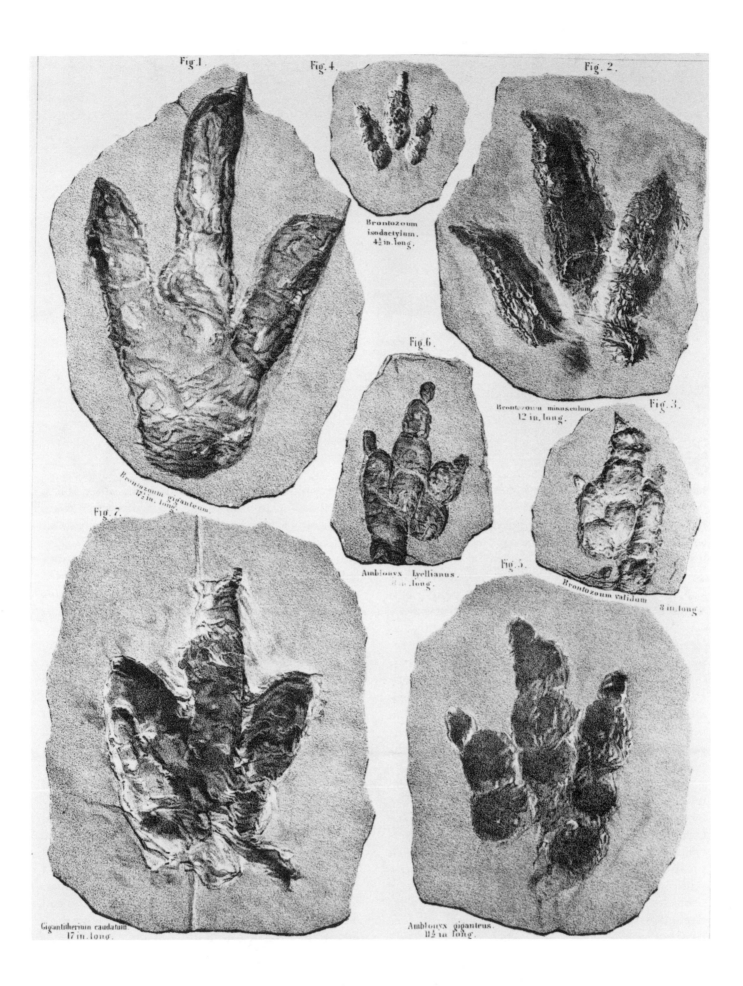

Fig. 1.

Fig. 4.

Fig. 2.

Brontozoum
isodactylum.
4½ in. long.

Fig. 6.

Brontozoum minusculum.
12 in. long.

Fig. 3.

Amblonyx Lyellianus.
8 in. long.

Fig. 5.

Brontozoum validum.
3 in. long.

Brontozoum giganteum.
17½ in. long.

Fig. 7.

Gigantitherium caudatum.
17 in. long.

Amblonyx giganteus.
11½ in. long.

3

Richard Owen's restoration of *Megalosaurus*, from *Geology and Inhabitants of the Ancient World* (London: Bradbury & Evans, 1854).

forelimbs and much larger hindlimbs. Leidy cautiously speculated that this creature, which he named *Hadrosaurus* and which clearly was related to the British *Iguanodon*, might have been bipedal at least part of the time.

Discoveries made in the late 1800's in Belgium and the American West provided compelling evidence that Leidy was right, and that many dinosaurs had walked on their hind legs. In 1878, at Bernissart, Belgium, coal miners discovered the remains of numerous individuals of *Iguanodon* in two underground galleries. These fossils provided very complete information about the skeleton of *Iguanodon*, and showed that, like *Hadrosaurus*, it had been at least a facultative biped. Owen's quadrupedal reconstruction of *Iguanodon*, based on incomplete remains, had been well off target.

At the same time as the excavations at Bernissart were being carried out, Othniel Charles Marsh of Yale University and Edward Drinker Cope of the Philadelphia Academy of Natural Sciences (with which Leidy also was affiliated) began prospecting for fossils in the western United States. Although initially friendly to one another, they soon developed a bitter rivalry. Both men were independently wealthy and both were brilliant scientists with colossal egos—and each was completely unscrupulous in his dealings with the other. They sniped at one another in the pages of learned journals, sent spies to snoop around each other's fossil sites, and in general behaved in a comically disgraceful manner—but they discovered dinosaurs galore. Their efforts made western North America *the* place to collect dinosaurs, and their discoveries demonstrated the great diversity of dinosaurs—creatures large and small, bipeds and quadrupeds, herbivores and carnivores.

In the latter part of the nineteenth century the study of dinosaurs was well-established: Cope and Marsh in the United States, Louis Dollo (who studied the iguanodonts from Bernissart), Thomas Henry Huxley, Harry Govier Seeley and Friedrich von Huene in Europe, and Florentino and Carlos Ameghino in South America, to name some of the more prominent workers.

In 1891 Henry Fairfield Osborn, a wealthy graduate of Princeton University and a close friend of Cope, founded the Department of Vertebrate Paleontology at the American Museum of Natural History in New York. Osborn gathered around himself an extremely able group of lieutenants, men like W. D. Matthew, Jacob Wortman (who had worked for Cope), and Walter Granger, who were distinguished paleontologists in their own right.

Still another of Osborn's associates was the central figure of R.T. Bird's memoir: Barnum Brown.

Barnum Brown was born in Carbondale, Kansas, on February 12, 1873, the youngest child of William and Clara Brown, who had settled in Kansas in the years before the Civil War. Supposedly the circus impresario P.T. Barnum took his show through the region at the time of Brown's birth, and the boy was named Barnum in the showman's honor. It was a fitting moniker; as will be apparent from Bird's story, Brown was endowed with a healthy dose of a showman's flamboyance. Brown's parents made a prosperous living by farming, digging and selling the coal that was so abundant in their area, and hauling freight among the U.S. Army outposts of the western interior.

At the age of sixteen Barnum was sent to Lawrence, Kansas to attend high school, and after that the University of Kansas. While in college he was very interested in archaeology and especially paleontology; he studied the latter under Samuel W. Williston, who had been a fossil collector for Marsh. Brown accompanied Williston on fossil collecting trips to Nebraska, South Dakota, and Wyoming. On one such trip he worked with a party from the American Museum led by Wortman, and made such a favorable impression that upon leaving the university he was hired by the museum (he finally received his degree in 1907).

In 1897 Osborn sent a group that included Wortman, Granger, and Brown to one of Marsh's old sites at Como Bluff, Wyoming. Marsh's collectors had found rich pickings of huge Jurassic dinosaurs and their diminutive mammalian contemporaries at Como Bluff, and Osborn hoped to obtain similar fossils for the American Museum. The following year an American Museum party discovered an extraordinarily rich site, the Bone Cabin Quarry, in the vicinity of Como Bluff. There were so many dinosaur bones here that a local sheepherder had used them to construct his home— hence the name of the site. Between 1898 and 1903 American Museum parties collected 483 dinosaur specimens at and in the vicinity of Bone Cabin Quarry, shipping them to New York in 275 crates that cumulatively weighed well over 73 tons. Among these specimens was the *Brontosaurus* skeleton that is mentioned frequently in Bird's memoir.

Other institutions were also actively searching for Jurassic dinosaurs in the American West at this time. Earl Douglass of Pittsburgh's Carnegie Museum of Natural History discovered a site near Vernal, Utah that proved to be even richer than the Bone Cabin Quarry. The Carnegie Museum worked this locality from 1909 to 1923; in 1915 it became Dinosaur National Monument. The Carnegie dinosaur collection was ultimately studied by Charles W. Gilmore of the Smithsonian Institution, a close friend of Barnum Brown.

The early twentieth century saw dinosaurs being collected and studied all over the world. In addition to the American expeditions for Jurassic dinosaurs already mentioned, Richard Swann Lull of Yale University studied the bones and footprints of dinosaurs and other animals of the Connecticut Valley, von Huene worked on Triassic dinosaurs in Germany and Brazil and Cretaceous dinosaurs in Patagonia (South America continues to yield many interesting finds in our day), Werner Janensch collected at a spectacular Jurassic site at Tendaguru (in what is today Tanzania), the deranged, politically ambitious nobleman Franz Nopsca studied Cretaceous dinosaurs in his native Transylvania, and Roy Chapman Andrews of the American Museum led an expedition to Central Asia, in the course of which the first dinosaur eggs known to science were found. (More recently Soviet, Polish, and Mongolian expeditions have continued work in Central Asia, finding a diversity of beautiful and often bizarre dinosaurs; the Chinese have in recent years made exciting discoveries in their country as well.)

As active as anyone at this time was Barnum Brown. Beginning in 1902 he worked in eastern Montana on Late Cretaceous beds of the Hell Creek Formation, collecting specimens of herbivorous dinosaurs related to *Iguanodon* and *Hadrosaurus* as well as skeletons of the huge carnivorous dinosaur *Tyrannosaurus*. In the midst of a personal tragedy, the death of Marion, his wife of six years, Brown left New York in 1910 to look for Late Cretaceous dinosaurs along the Red Deer River of Alberta.

Canadian geologists had discovered dinosaur bones in this region in the late 1800's, and these bones had been seen by Cope and Osborn, so Brown knew that his chances of finding additional fossil material were good. He was very successful, both in 1910 and 1911—in fact, he collected so many dinosaur specimens that the Canadians became alarmed at how many fossils he was removing from their territory. The Canadian Geological Survey hired an American fossil hunter, Charles H. Sternberg (who had collected for Cope), and his sons George, Charles M., and Levi, to collect dinosaurs for Canadian institutions, and they competed with Brown for fossils along the Red Deer from 1912 to 1915. In contrast to the rancor that had characterized the Cope-Marsh rivalry, competition between Brown and the Sternbergs was friendly, and both parties collected an unprecedented diversity and abundance of dinosaurs. Brown's operation along the Red Deer ended in 1915, as did the team effort of the Sternbergs. However, individual members of the Sternberg family continued exploring the region for fossils for many years. In particular, Levi collected for the Royal Ontario Museum, and Charles M. for the National Museums of Canada. The Red Deer Valley is still a rich fossil field, particularly in what is today Dinosaur Provincial Park, where field crews from the Tyrrell Museum of Palaeontology in Drumheller carry out an intensive program of collection and research.

Upon completion of his Canadian work Brown turned his attention to fossil vertebrates in many other parts of the world, travelling to Cuba, Mexico, India, Pakistan, and the Mediterranean region. While collecting fossil mammals on the Indian Subcontinent, where he worked from 1922 to 1924, he married his second wife, Lilian (the Mrs. Brown of Bird's memoir), who eventually wrote three books describing fossil hunting trips on which she had accompanied her husband: *I Married a Dinosaur, Cleopatra Slept Here*, and *Bring 'em Back Petrified*. Barnum Brown was made Curator of Vertebrate Paleontology in 1927, so he was at the height of his career as R.T. Bird's account begins. Although Brown had begun graduate studies at Columbia University in 1897, the press of museum work forced him to abandon them, and he never did earn a formal postgraduate degree; however, Lehigh University awarded him an honorary Doctor of Science degree in 1934.

In many ways Brown was the one of the last and greatest of the old-time dinosaur hunters. In his time the main effort of field work was to collect new or well-preserved specimens; paleontologists were primarily interested in finding new faunas and working out the evolutionary relationships of the creatures they studied. While these goals are still important in fossil hunting, there is an added dimension to field work nowadays. Modern paleontologists are also concerned with questions of how fossil assemblages formed and how they can be used to reconstruct the paleoecology of ancient organisms. Modern field collecting techniques are consequently more sophisticated and quantitative than those of Brown's day. For studies of this kind, poorly-preserved and fragmentary specimens (which were often ignored in the pioneer era of vertebrate paleontology) can be just as important as the most complete and exquisitely preserved skeleton. This is not to say that the old-timers like Brown were uninterested in matters of paleoecology and the formation of fossil assemblages—they were indeed interested, but such questions were of secondary importance to them.

At one point R.T. Bird refers to Brown as the last and greatest of the "three kings" of the dinosaurs, Cope and Marsh being the other members of this paleontological triumvirate. Bird's loyalty to his boss is appealing, but it should be apparent that setting Marsh, Cope, and Brown (despite their unquestioned contributions) above all others is a bit of an affront to the many other workers who also had significant accomplishments in the study of dinosaurs.

Barnum Brown spent sixty-six years working for the American Museum. In 1942 he left his post as curator because he had reached the mandatory age of retirement. During World War II he was a consultant to the U.S. Government, working for the Office of Strategic Services and the Board of Economic Warfare on matters related to petroleum exploration (he had been an adviser to the government and the petroleum industry during World War I as well, and periodically to petroleum companies between the wars). He continued to work as a petroleum geologist, and to collect fossils, after the Second World War. He died in 1963, a week short of his ninetieth birthday.

Geology has its own peculiar jargon (just like any other field of human endeavor), and an understanding of this terminology is necessary for full appreciation of Bird's memoir. The subdivision of geology concerned with deciphering earth history from rock layers and their contained fossils is stratigraphy. For the most part, stratigraphy is concerned with the study of sedimentary rocks—rocks that form from the consolidation of sediments. Some sediments form from the weathering and disintegration of pre-existing rocks, such degradation yielding particles of clay, silt, and sand size, or pebbles and cobbles of even larger size. Sedimentary particles are transported by wind and water to low-lying areas (particularly on the floors of seas and oceans) where they accumulate as sedimentary layers.

As the cumulative thickness of these layers increases, their weight compresses the sedimentary mass, squeezing out water and squashing sedimentary particles together; at the same time, lime, silica, or other cements may be precipitated in the spaces among the particles. Compaction and cementation thus convert sediments to sedimentary rocks. A sedimentary rock composed of particles of clay size is called a shale; of silt size, a siltstone; of sand size, a sandstone; and of still larger particles, a conglomerate.

Not all sediments form from the destruction of pre-existing rocks, however; some are generated by chemical and especially biological processes. The prime example of the latter is limestone. Many marine organisms (algae, certain protozoans, and many animals) secrete skeletons or microscopic needles of calcium carbonate (lime), and these limy skeletons either consolidate in place (as in reefs), or accumulate postmortem as sediments which later become sedimentary rocks by processes similar to those responsible for the conversion of sediments to rock in shales, siltstones, sandstones, and conglomerates.

When a geologist goes into the field to describe the sedimentary rocks he or she sees, the basic stratigraphic unit defined is the formation. A formation is a sequence of rock layers that formed by the same process(es), and which differs in some distinctive feature(s) such as its kind of sedimentary rock, or the texture (e.g., coarse-grained or fine-grained) of its sedimentary particles, from other formations above and below it. Recognition of a formation is somewhat subjective—almost as much an art as a science.

Typically a formation is named for some geographic feature in the region where it is first recognized; thus the Morrison Formation and the Cloverly Formation, prominent rock units in Bird's story, were named for Morrison, Colorado and the old Cloverly Post Office near Shell, Wyoming, respectively. Where a formation is mainly composed of a particular kind of rock, its name may recognize that fact, as in the Navajo Sandstone, the Mancos Shale, and the Glen Rose Limestone (also called the Glen Rose Formation, but by either name an important formation in Bird's account).

Formations are descriptive units and have nothing directly to do with the concept of geologic age. A geographically widespread formation may vary in age over the region in which it occurs, and two rock sequences that formed at the same time will not for that reason be considered parts of the same formation.

However, stratigraphy does have rock units that do reflect geologic age; such rock units are based on the succession of different kinds of fossils through geologic time. A zone is a group of rock layers defined by the fossils they contain. One kind of zone comprises all of the sedimentary layers that were deposited during the existence of a particular kind of fossil. Another kind represents the stratigraphic interval over which the geologic ranges of several different kinds of fossils overlap. Other types of zones are based on the local co-occurrence of several different types of fossils, or on the stratigraphic interval over which a particular kind of fossil is especially abundant.

A succession of zones constitutes a more inclusive stratigraphic unit, a stage; several stages together make up a series, and a sequence of series, a system. Geologists refer to these units as time-rock units. Geologists also recognize intervals of time that correspond to these time-rock units. Thus the Cretaceous Period represents the interval of time during which the rocks of the Cretaceous System formed. Several such periods are grouped together to constitute an era (some geologists recognize a time-rock unit, the erathem, that corresponds to the era).

From the standpoint of R. T. Bird and Barnum Brown, the Mesozoic Era with its three periods, the Triassic, Jurassic, and Cretaceous, represents the geologic interval of greatest interest. This was the Age of Reptiles.

Like stratigraphy, the study of dinosaurs has its own complex terminology, at least the rudiments of which must be grasped in order to understand Bird's account of his career. When a paleontologist describes a new kind of fossil, he or she assigns it a scientific name, so that workers around the world will be able to refer to the same creature and be sure that their colleagues know which beast is being discussed. A scientific name has two components, a generic name and a specific name. Thus the scientific name for modern human beings is *Homo sapiens*; by convention, the first letter of the generic name is capitalized, and both the generic and specific names are either underlined or itali-

Era	Period		Date of Beginning of Period (Millions of Years Ago)	Significant Events
CENOZOIC	Neogene		25	Modern mammal groups dominant. Appearance of humans late in period. Uplift in western North America to produce modern topography. Widespread mountain-building around world. Deterioration of climate culminating in episodes of glaciation.
CENOZOIC	Paleogene		65	Archaic mammals and flowering plants dominant on land. Continued mountain-building in western North America.
MESOZOIC	Cretaceous		135	Peak diversity of dinosaurs. Flowering plants became dominant. Modern groups of fishes and marine invertebrates. Mass extinction at end of period. Mountain-building and periodic flooding in western North America. Continents approaching their modern positions.
MESOZOIC	Jurassic		192	Dinosaurs dominant. First birds. Flooding of western North America, followed by mountain-building and restoration of terrestrial conditions. Pangaea breaking up.
MESOZOIC	Triassic		230	Widespread coniferous forests. Diversity of therapsids and thecodonts. First dinosaurs and mammals. Mountain-building in western North America. Widespread deposition of red-colored sediments in North America.
PALEOZOIC	Permian		290	Continental collisions to produce Pangaea. Continued mountain-building in Southwest and Appalachian region. Mass extinction of marine animals at end of period. Diversity of reptiles. Widespread glaciation in Southern Hemisphere.
PALEOZOIC	Pennsylvanian	Carboniferous	320	Non-marine environments in eastern North America. Widespread occurrence of coal swamp forests. Diversity of insects and amphibians. First reptiles. Mountain-building in southwestern North America and Appalachian region.
PALEOZOIC	Mississippian	Carboniferous	350	Last major episode of limestone deposition in North America. Diversity of amphibians.
PALEOZOIC	Devonian		410	Continued flooding of North America and widespread reef development. Diversity of fishes. First amphibians. Development of trees and forests. Mountain-building in the Appalachian region, Alaska, and Arctic Canada.
PALEOZOIC	Silurian		435	Continued flooding of North America. Widespread occurrence of reefs. Appearance of fishes with true jaws. First land plants and invertebrates. Mountain-building in northern and northeastern Canada.
PALEOZOIC	Ordovician		485	Jawless fishes. Continued flooding of North America, with limestone deposition. Mountain-building in the Appalachian region.
PALEOZOIC	Cambrian		560	Widespread acquisition of hard skeletons by invertebrate animals. Extensive flooding of North America; deposition of quartz sands.

PRECAMBRIAN   3800+

Origin of earth, atmosphere, and oceans. Significant episodes of mountain-building, continental glaciation. Origin of life. Primitive multicellular plants and animals.

Formations, Faunas, or Sites Mentioned in Bird's Memoir

Bridger Formation

Late Cretaceous: Hell Creek Formation, Mesaverde Formation, Judith River Formation, Aguja Formation

Early Cretaceous: Cloverly Formation, Glen Rose Formation

Late Jurassic: Morrison Formation (Howe Quarry, Dinosaur National Monument, Como Bluff, Bone Cabin Quarry)

Early Jurassic: Dinosaur track sites in Connecticut River Valley and northern Arizona

Late Triassic: Chinle Formation (Petrified Forest)

Amphibian tracks in coal mine, Wilkes-Barre, Pennsylvania

The geological time scale, showing the ages of the faunas, formations, or sites described in Bird's account. Note that many geologists subdivide the Cenozoic Era into Tertiary and Quaternary Periods (the terminology Bird uses) instead of Paleogene and Neogene. The Tertiary corresponds to the Paleogene plus most of the Neogene, while the Quaternary designates the last few million years of the Neogene. Compiled by James O. Farlow.

cized. There is commonly more than one species in a genus. In doglike mammals, for example, there are wolves (*Canis lupus*), coyotes (*Canis latrans*), and jackals (such as *Canis aureus*, *Canis adustus*, and *Canis mesomelas*), to name a few. These animals are sufficiently distinct to deserve their own specific names, but related closely enough to one another to be placed in the same genus.

The same principles are used for dinosaurs. For example, in his classic monograph on armored dinosaurs, C. W. Gilmore recognized nine species of the genus *Stegosaurus*: *S. armatus*, *S. discurus*, *S. durobrivensis*, *S. longispinus*, *S. priscus*, *S. seeleyanus*, *S. stenops*, *S. sulcatus*, and *S. ungulatus*. Similarly, in his review of ceratopsian dinosaurs Lull recognized nine species of *Triceratops*.

Although in theory identifying different species of a particular genus is a valid procedure, in practice the results for dinosaurs have been less than gratifying from the standpoint of modern paleontologists. Older workers recognized new species on the basis of what today seem flimsy grounds, and one can reasonably doubt that there really were nine distinct species of *Stegosaurus* or *Triceratops*. Much current work in classifying dinosaurs is consequently concerned with reducing the number of recognized species (in the jargon, "synonymizing" them). In a recent study of hadrosaurs from the Judith River Formation, for example, the number of species of *Corythosaurus* was reduced from six to one and the number of species of *Lambeosaurus* from three to two.

In part due to the uncertain validity of named species and in part out of convenience, when referring to dinosaurs paleontologists usually give the generic name and drop the specific name, except in technical articles specifically concerned with classification. The same is done in the popular literature, with one glaring exception: for reasons that I don't understand, children's books and other popular publications almost invariably refer to *Tyrannosaurus rex* rather than *Tyrannosaurus*.

Just as there may be more than one species in a genus, several related genera can be grouped into a family. Thus the family Tyrannosauridae contains such genera of carnivorous dinosaurs as *Albertosaurus*, *Daspletosaurus*, *Tarbosaurus*, and *Tyrannosaurus* itself. Paleontologists often informally refer to these creatures as tyrannosaurs.

Family names and informal names, unlike formal generic and specific names, are not italicized—another convention. Thus one italicizes and capitalizes the generic name *Iguanodon*, but not the informal name iguanodont, which can refer to *Iguanodon* itself or to related dinosaurs. Similarly, Bird uses the term brontosaur informally on occasion when talking about the genus *Brontosaurus* (a generic name which, despite its familiarity, is considered by most paleontologists to be a synonym of *Apatosaurus*, a name which was coined earlier and so has priority over *Brontosaurus*). Other workers sometimes use brontosaur as an informal name for all of the sauropod dinosaurs.

The work of Barnum Brown and his fellow students of dinosaurs laid the foundation for our present understanding of the great reptiles. Although Owen, limited as he was by the fragmentary skeletal material available to him, believed dinosaurs to have comprised a single group of reptiles, we now know that there were two basic groups of dinosaurs, distinguished from each other by the construction of their hip skeletons. Saurischian dinosaurs had the three bones of the hip, the ilium, ischium, and pubis, constructed in such a way as jointly to form a triangular shape in side view. In ornithischian dinosaurs, on the other hand, the hip skeleton had a more rectangular appearance, superficially similar to that of birds (ironically, birds appear to be more closely related to saurischians, however). Some paleontologists believe that ornithischians evolved from primitive saurischians, but the majority view is that the two groups arose independently from a group of reptiles known as thecodonts.

Dinosaurs made their appearance during the latter part of the Triassic Period, in a world very different from that in which we live today. The various continents of our modern world were then conjoined to form a huge supercontinent known to geologists as Pangaea. This supercontinent had come into existence as pre-existing continental fragments had drifted together during the middle and later parts of the Paleozoic Era. However, Pangaea had no sooner come into being than it began to break up, eventually splitting apart to form the continents that we know today. The breakup of Pangaea began in the Triassic, with extensive rift valleys forming in Europe, North America, South America, and northwestern Africa, accompanied by volcanic eruptions.

Late Triassic forests were dominated by a variety of conifers, cycads, and ferns, and roaming through these forests were a diversity of reptiles and amphibians. The dominant large reptiles were therapsids, creatures which structurally foreshadowed mammals to varying degrees—indeed, mammals appeared during the Late Triassic, derived from certain small, advanced carnivorous therapsids. In addition to the hordes of herbivorous and carnivorous therapsids, there were large lizard-like beasts with peculiar parrot-like beaks, known as rhynchosaurs. Large amphibians similar to the animal found by Bird in Arizona swam in the rivers, lakes, and ponds. Contesting the dominance of the therapsids were the thecodonts, a diverse group of reptiles that included large and small, quadrupedal and bipedal, and herbivorous as well as carnivorous forms. Among the thecodonts were large, crocodile-like creatures known as phytosaurs—the skeleton of one of which Brown and Bird would find in northern Arizona. Less spectacular inhabitants of the Triassic world included early turtles and lizards.

Derived from advanced thecodonts were the earliest dinosaurs. Most of these were carnivorous saurischians known as theropods, and most of these theropods were coelurosaurs—slimly built creatures with long legs, tails, and necks, and jaws armed with sharp teeth. Most coelurosaurs were of modest size—ostrich-sized or smaller—but larger dinosaurs were also present in the Triassic fauna. More ponderous theropods known as carnosaurs, able by the end of the Triassic to match the toughest therapsid or thecodont claw for claw and tooth for tooth, were also making their appearance, as were equally large, long-necked, herbivorous saurischians known as prosauropods. Ornithischians were also present, represented by relatively small, slimly built herbivores. As the Triassic Period ended and the Jurassic Period began, dinosaurs left footprints in countless thousands along the muddy shores of lakes and rivers in the rift valleys of eastern North America and in northern Arizona, tracks that would eventually delight the likes of Edward Hitchcock, R.S. Lull, Barnum Brown, and R.T. Bird.

By the Late Jurassic, Pangaea had separated into northern and southern portions, but it was still possible for land animals periodically to move between these portions. As a result the land faunas were still fairly similar in all parts of the world. Thecodonts and therapsids had fallen by the evolutionary wayside, leaving dinosaurs as the dominant large land animals—and "large" fails to do justice to many of the Late Jurassic dinosaurs. Modest-sized coelurosaurs and bipedal ornithischians (ornithopods) there still were, but the carnosaurs had become huge bipedal dragons, and the prosauropods or creatures like them had given rise to the even more titanic sauropods, quadrupedal giants with long necks to feed from the crowns of tall trees. Stegosaurs—large, ungainly, quadrupedal, herbivorous ornithischians adorned with bony plates and sharp spikes—made their prickly presence known in many parts of the world.

In the western United States, the terrestrial conditions (except in the extreme West, which was marine) of the Late Triassic and Early Jurassic had given way to marine environments in the Middle Jurassic, with a shallow seaway flooding the continent from the Arctic to the Gulf region. In the latest Jurassic, however, mountain building caused

Scutellosaurus, a primitive ornithischian from Early Jurassic rocks of northern Arizona. Restoration by Margaret Colbert.

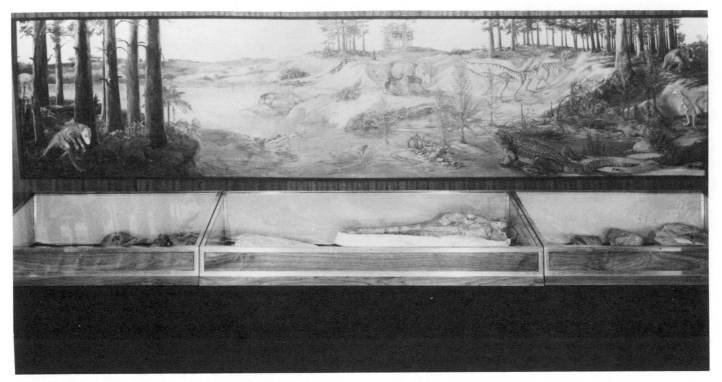

Animals of the Chinle Formation (Late Triassic) of northern Arizona. In the foreground are two phytosaurs, one of which dines on a lungfish while the other guards its newly-hatched young. Between the phytosaurs swims *Metoposaurus*, an amphibian related to the creature whose skull Bird found in Arizona. A bipedal carnivorous thecodont (*Hesperosuchus*) eats an unwary hatchling phytosaur at the far right, and another *Hesperosuchus* is at the far left foreground of the picture.

Two quadrupedal, armored thecodonts (probably herbivorous), *Desmatosuchus* (at the water's edge) and *Typothorax* (above and to the right of *Desmatosuchus*) occupy the middle of the scene; to their right is the tusked, herbivorous therapsid, *Placerias*. Scampering up the hill are three coelurosaurs (*Coelophysis*). Painting by Margaret Colbert; used by permission of the National Park Service.

the sea to drain away, restoring terrestrial conditions over much of the West. Sediments eroded from these mountains were carried eastward by rivers and deposited over a vast area in what would become the Morrison Formation, a mother lode of dinosaur bones that would one day be mined at Como Bluff, the Bone Cabin Quarry, Howe Quarry, and Dinosaur National Monument.

You can't keep a good seaway down, however, and during the Cretaceous Period North America was once again split into eastern and western portions by marine environments in the midcontinent. This sea waxed and waned in extent, with terrestrial environments encroaching repeatedly from the west to form such units as the Mesaverde Formation (where Bird collected the footprints of a huge ornithopod from the roof of the States Mine), the Judith River (or Oldman) Formation (the scene of the friendly rivalry between Barnum Brown and the Sternbergs in Alberta), and the Hell Creek Formation (where Brown found *Tyrannosaurus*). By Hell Creek time the transcontinental seaway was re-

treating for good, and it dried up completely during the early Cenozoic.

Extensive flooding of the continents typified the Cretaceous in all parts of the world, but aside from this world geography was taking on an increasingly familiar appearance. North and South America had separated from Africa to create the Atlantic Ocean, although North America remained attached to Eurasia into the early Cenozoic. India had separated from Africa and begun drifting toward an eventual head-on collision with Eurasia, the dents from which would form the Himalayan Mountains. A conjoined Australia and Antarctica had split away from Africa and were soon to part company with each other.

Plant and animal life of the Cretaceous comprised a mixture of the familiar and the exotic. By the end of the period the marine invertebrates and fishes were represented by groups we see today. On land, the flowering plants appeared during the Late Jurassic or Early Cretaceous, and by the end of the Cretaceous had replaced conifers as the dominant land plants. Many of the modern insect groups evolved in association with the flowering plants. Birds had appeared in the Jurassic, and in Cretaceous times seem to have been moderately diverse, perhaps giving the pterosaurs, or flying reptiles (animals related to the dinosaurs), a bit of a hard time. Crocodiles (some quite large), turtles, and lizards were widespread and abundant (all three groups had initially appeared in the Triassic), as were frogs and salamanders, and a few snakes showed up by the Late Cretaceous.

So much for the familiar. With regard to the exotic, ammonoids, shelled molluscs related to modern squids and octopuses, remained abundant in Cretaceous seas, as they had earlier in the Mesozoic. The same was true of a host of large marine reptiles (which, like pterosaurs, were not dinosaurs, although the general public usually thinks of them as such).

The dinosaurs were at their zenith. Sauropods remained important, leaving their tracks in what would become Texas and Arkansas, although they were no longer abundant at some of the places in North America where they'd been common earlier. Carnosaurs were perhaps bigger and nastier than ever, and smaller theropods were present in bewildering diversity, some of them having unusually large brains by reptilian standards.

Most stegosaurs did not last much beyond the end of the Jurassic; they were replaced by many new groups of ornithischians, perhaps evolved in response to new food sources created by the spread of the flowering plants. Iguanodonts (including *Tenontosaurus*, the Cloverly form mentioned by Bird) and hypsilophodonts were primitive ornithopods that were important in the Early Cretaceous, *Iguanodon* itself being the dinosaur Mantell had described. By the Late Cretaceous the most abundant large dinosaurs in many places were advanced ornithopods, hadrosaurs, facultative bipeds with a complex dentition for processing plant material. Many hadrosaurs had bizarre hollow crests on their heads, perhaps used to attract mates and cow rivals during the mating season. There is evidence that some of these hollow crests may have had resonating properties similar to those of certain Medieval wind instruments; this would have permitted crested hadrosaurs to utter more impressive honks or roars. At least some hadrosaurs and other ornithopods are known to have laid eggs at communal nesting sites to which they returned year after year, behavior similar to that seen in some modern crocodiles. Pachycephalosaurs were another group of Cretaceous ornithopods. With their skulls encased in a thick covering of bone, these dinosaurs may have butted their heads together during mating season rivalries.

A second major group of Cretaceous ornithischians was the ankylosaurs. Encased in a cuirass of bone and variously protected by a row of spines

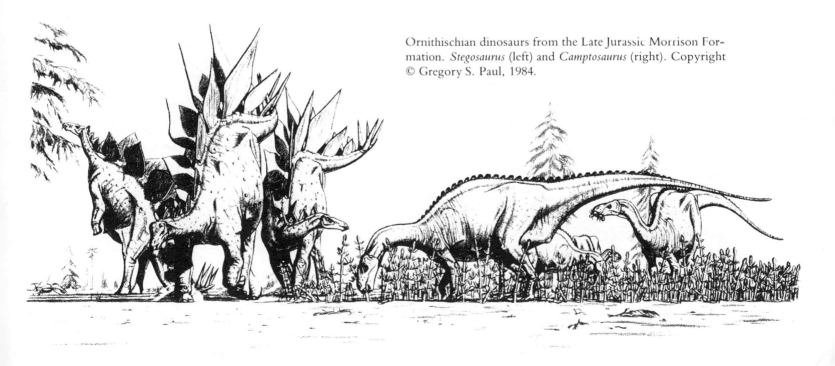

Ornithischian dinosaurs from the Late Jurassic Morrison Formation. *Stegosaurus* (left) and *Camptosaurus* (right). Copyright © Gregory S. Paul, 1984.

Sauropod dinosaurs from the Late Jurassic Morrison Formation of the western United States. Left to right: *Camarasaurus*, two *Barosaurus*, and *Apatosaurus* (*Brontosaurus*). Copyright © Gregory S. Paul, 1984.

projecting outward from the sides of the body, or a massive tail club, these herbivores were well equipped to resist the attacks of most carnivores.

Rivalling the hadrosaurs in exotic headgear were the ceratopsians, or horned dinosaurs. These reptiles were characterized by a shearing dentition and large bony frills projecting from the backs of their skulls. The frill probably served as an attachment site for powerful jaw muscles, a counterbalance for the massive skull, and as a display structure similar to the crests of hadrosaurs. Some ceratopsians had large horns above their eyes and a smaller horn on the snout, some reversed this, and some got by with a large bony boss instead of horns. Presumably these differences in cranial armament reflect differences among ceratopsian species in combat behavior during territorial contests or dominance disputes and fights over mating rights, just as similar differences in horn shape go along with differences in fighting styles in modern antelopes. The mass occurrence of ceratopsians of the same species, represented by individuals of varying size in some fossil assemblages, suggests that at least some horned dinosaurs lived (and died during sudden local catastrophes) in herds.

And then the Cretaceous ended, and all of the dinosaurs were gone—and with them the assortment of other terrestrial and marine organisms that had helped to give the Mesozoic Era its distinctive flavor. They perished in a mass extinction, just as an even greater episode of extinction had punctuated the end of the Paleozoic Era. Such mass extinctions have been a recurrent feature of earth history, and in recent years have received much attention from scientists representing a number of disciplines. Some scientists believe that such disasters have extraterrestrial causes, such as collisions of large asteroids with the earth. Other workers prefer terrestrial explanations, such as changes in sea level and their accompanying effects on climate and living space for organisms. Whatever the cause, the extinction of the dinosaurs made it possible for the tiny Mesozoic mammals to replace them, evolving into the diversity of large and small animals that typified the Cenozoic Era down to the present day.

James O. Farlow
Indiana University-Purdue University
at Fort Wayne

Dinosaurs from the Late Cretaceous Judith River (Oldman) Formation of Western Canada. A herd of ceratopsians (*Styracosaurus*) warily observes a carnivorous dinosaur (*Albertosaurus*). Restoration by Gregory S. Paul. Courtesy Sylvia and Steve Massey-Czerkas.

15

Painting of *Saurolophus*, a Late Cretaceous hadrosaur, by Eleanor M. Kish from *A Vanished World: The Dinosaurs of Western Canada* by Dale A. Russell (Ottawa, 1977). Reproduced courtesy of the National Museum of Natural Sciences, National Museums of Canada.

Painting of *Hypacrosaurus*, a Late Cretaceous hadrosaur, by Eleanor M. Kish from *A Vanished World: The Dinosaurs of Western Canada* by Dale A. Russell (Ottawa, 1977). Reproduced courtesy of the National Museum of Natural Sciences, National Museums of Canada.

Prologue

WHEN ROLAND T. BIRD was a boy, there was a great deal more agreement about what dinosaurs were really like than there is today. Everyone believed they were all cold-blooded, because they belonged to the Reptilia, and scientists knew, of course, that all reptiles are cold-blooded. Everyone knew that a tremendous creature like the brontosaur could not possibly support its great weight on land; therefore he must have spent his time almost completely in the water.

Rebels among the scientists have brought changes to this neat and faulty picture in recent times. Roland T. Bird was probably the first rebel to make an impression on the study of dinosaurs. He is on record as the first to take the brontosaurs out of the water and set their feet on solid ground. His studies and his draftsmanship on the tracks at Bandera and in the Paluxy River in Texas were the first evidence in the world that the great quadrupeds came out of the water from time to time, that they traveled in herds, and that they gave some protection to their young. More recent work indicates that Bird only scratched the surface of the story but made the first important discoveries. Later scientists, inspired in part by his work, have proposed that dinosaurs were physiologically and behaviorally advanced over living reptiles, and a few would remove these creatures from the class Reptilia altogether.

Bird began life as a lonesome cowboy, riding from job to job on a rusty Harley motorcycle with a cumbersome fold-away trailer where the sidecar should be. A drop-out from junior high school, a loner set back by a bout with rheumatic fever when a child, he was a poor bet for leaving his mark on history or scientific knowledge.

A lucky find of a fossil jaw, the jaw of a previously unknown amphibian, gave him the key to success, to achievement in the field of paleontology. It was the key to the office of Barnum Brown, a leader in this field, the man who had found for Harry Sinclair the brontosaur his oil company rode as their emblem for forty years, a man who had contributed as much to the study of dinosaurs as anyone.

Impressed no more by the strange jaw than by Bird's eager attitude, Brown gave him a pick and shovel on a new dig north of Shell, Wyoming. Among the multitude of bones were the remains of thirteen brontosaurs. Bird, first with his pick and shovel, then with his whiskbroom and awl, worked his way here into Brown's interests and his heart. He didn't do this all on his own; Brown gave him guidance no millionaire's son could have bought in any university.

When his health broke at the end of World War II and he had to leave the job, Bird regarded himself as professionally dead. His own story, told by himself, ends with what he thought to be the last deed of his that mattered.

Happily for himself and for us, he was reborn by a letter from Dr. Edwin Colbert, who shared the front rank of the dinosaur people with Brown. He offered Bird a chance to come back to the American Museum of Natural History in New York and put together the broken pieces of brontosaur tracks which Bird had shipped from the Paluxy River to the museum fourteen years earlier, before the war. If Bird could make no sense of the contents of broken boxes, the lost labels and the cracked plaster that had lain out in the weather for fourteen years, the tracks might be of no scientific value. Colbert guessed the job might take a month, maybe a bit more. It took six months. It is the postscript to Bird's own record of his life.

A swimming iguanodont. Such animals left tracks in the Early Cretaceous deposits of British Columbia, Texas, and other parts of the world. Restoration by Gregory S. Paul.

Courtesy Tyrrell Museum of Palaeontology, Drumheller, Alberta.

Painting of *Dromiceiomimus*, a Late Cretaceous coelurosaur, by Eleanor M. Kish from *A Vanished World: The Dinosaurs of Western Canada* by Dale A. Russell (Ottawa, 1977). Re-produced courtesy of the National Museum of Natural Sciences, National Museums of Canada.

I

AT FIRST the distant buttes were purple, little purple boxes floating on the Arizona vastness. The desert road stretched long and straight and empty toward the nearest of the box-like hills. I checked the cycle's mileage on the speedometer against the road map. Winslow and Holbrook and much of the Painted Desert were well behind me. Seventy miles yet to Pine, and the chill November day in 1932 was nearly done. If I could find a place to build a campfire at the foot of one of those box-walls ahead . . .

The trail was rough and rutted; the biggest bumps jarred the machine to the limits of its rugged bolts. Dust, caught in updraft behind the windshield, eddied into my face. I took a fresh grip on my bouncing body and the Harley's handlebars, sifted out the sand between my teeth for a bite of fresh air . . . and ran out of dust. The road ran over a rocky outcrop and dived down a rocky draw. I fought to hang on to the bucking machine. The draw, of course, cut off my view of the purple buttes. We crossed a dry wash, climbed again to level ground. Propinquity and the paling sun had changed the purple buttes to brownish red. As the Harley and I drew closer, the brownish reds became bands of red and brown, the colors less vivid than those of the buttes on the Painted Desert east of Holbrook.

The Harley and I waddled gracelessly down another draw, but the buttes were too close now to be lost to sight. The bands of clay and sandstone resembled layer cakes, varied in area but all exactly the same height. Nature, in cutting the desert to its present level, had sought somehow to preserve in these abutments remnants of an ancient horizon. I had first become aware of this in the Petrified Forest, but where the vast plain had started and where it would end I did not know. It had occurred to me as far back as Holbrook that I might follow the march of the box-like hills west and . . . somewhere beyond the limits of the strictly patrolled government park . . . I might find a petrified log of my very own, since the lay of the land was much the same.

A clay apron reached out from the base of the nearest butte; it was from such clays as these that the logs of the Petrified Forest had weathered out. I could still see in mind's eye the great boles lying everywhere, scattered like logs in an old mill pond, truly an Enchanted Forest, showing jasper and quartz and agate interiors, flashing colors of a shattered rainbow. "These trees," the park ranger said, casually as if talking about breakfast, "were alive around two hundred million years ago. They grew away back in the Age of Reptiles . . . the days of the dinosaurs."

I turned the machine toward the nearest butte, crashing through creosote bush and sage. The beat of stems under the sidecar sounded like splashing water. The machine with its heavy sidecar groaned and strained. I cut the switch, dismounted, and started walking toward the base of the butte. It was only some forty feet high but offered a good haven out of the rising night wind.

To think that dinosaurs had known the world I could see cross-sectioned in the faces of the weathered buttes! I knew little enough about the great beasts, but the ranger's words . . . I called to mind the great skeletons in New York's American Museum of Natural History which I had known as a boy; then they weren't associated with anything in the land of the living. But I was *here* now . . . where *they had* been! I looked about curiously, half expecting to find bones lying about.

The bands of clay and rock that made up the buttes were capped with coarse sandstone, several

Deinonychus, a medium-sized carnivorous dinosaur from the Early Cretaceous Cloverly Formation of Montana. Restoration by Robert T. Bakker. Courtesy Peabody Museum of Natural History.

feet in thickness, emphasizing the layer cake impression. The ranger at the Petrified Forest had explained that this had once been a great marine basin into which silt and sand and mud had been dumped by mighty rivers. Trees, perhaps uprooted by storm, floated into this basin to become waterlogged and buried.

Flat slabs of red sandstone, thin as boards, lay scattered through the soft clay of the base. Their undersides undulated every inch or so, varying the thickness. I picked up a piece, perhaps a two-foot slab, smooth as a slate on one side, marked on the other by rhythmic undulations, like the record left in shore sand by a retreating tide.

A retreating tide! Memory flooded, and I was a boy again in Rye, New York, on Long Island Sound, and the tide had just gone out, leaving a ripple-marked sandy beach. Here in my hand was just such a bit of beach, frozen in stone, a record of a lazy day two hundred million years gone when dinosaurs had roamed.

Ahead lay a broad exposure, unbroken, its expanse of ripple marks etched sharply by the last glimmers of sunlight, running up into the cover of the butte wall. Sometimes, in Rye, the ripples were smudged by the bare feet of bathers. I looked about me, trying to picture an ancient shoreline. Or perhaps the ripple marks were made on a sandy flat in a river bed. In any case, it had been a warmer world,

a fine world for great reptilians. If these ripple marks were smudged anywhere, it would not have been by barefoot bathers.

I moved about, inspecting with care the unbroken expanse, turning over some of the broken bits, covering the terrain with care until it grew too dark to see.

As I turned to give up for the night, a thin sliver of gold from under a low-hanging cloud conspired with reflection from the edge of the butte to give me a minute's respite. In this bit of time, a slab of rock tipped up from the shadows. The dark outline of a crocodile-like mouth printed in stone seemed to reach up in the dim light for my ankle. My heart skipped a beat, sputtered, went on. I stumbled, recovered my balance, and the stone mouth dropped back into its pool of liquid darkness, clacking.

The form in deepest shadow made no further movement. I stepped forward and bent over the object in the wash, lifting it into the last of the waning twilight. Running my fingers around a long dark smear across the surface of the stone, I felt the edge of a mouth-like form, the prints of small, sharp, numerous teeth. It was some sort of a fossil skull.

The rock was not too large nor too heavy to lift. I tugged it to the bank of the wash, where the light was slightly better. Only the head . . . or rather, the skull . . . was represented on the surface of the rock; the marginal outline, the rows of small sharp teeth, the oval openings into a long snout . . . all were perfectly shown in deep impression. Of actual bone there was nothing left, of course. The breaks across the stone seemed to be clean and fresh. When the rest of this slab had broken away, it had carried with it not only the bone but the top half of the impression. What lay before me was a palate imprint where the upper part of a large mouth had rested, a long time ago.

I turned again to the wash. It was littered everywhere with broken bits of flat rock, little rocks, big rocks, little slabs, big slabs, a long section that had slumped from the butte's capping. Surely the other half of the strange skull imprint was here somewhere. Even the skull itself, perhaps even more of the creature's skeleton. The surface of the nearest sizable slab was barren of marks. I turned it over . . . nothing on the other side. I followed the wash up to the capping and down as far as rocks of any size had been carried. No fossil bones; no imprints. It was full dark now, and fingers aren't a dependable substitute for eyes. The rising night wind began to whisper mournfully through the sagebrush, as if it too sought for what it could not find. "I've just missed it because I can't see," I thought. "I'll find it, come morning."

Turning west out of the wash, I made my way toward the dim shape of the Harley, lit only by starlight now. On the sidecar chassis was a small one-man camp trailer of my own building, folded down for daytime travel. I switched on the light and drove the outfit toward the wash, to a slightly more open spot. With the headlamp to see by, I pulled dead sagebrush for a campfire. The gnarled, knotty wood was fat, and splinters kindled readily. With the fire lighting up the area I had picked for a campsite, I turned my attention to setting up. In only moments, I had unfolded and rigged my little camper and uncovered a comfortable mattress.

Supper came from the contents of my tiny pantry, and the brushwood fire soon burned down, making a fine bed for roast sweet potatoes. Fresh meat went into the skillet and onto the bed of coals. When the cooking was done, I heaped the balance of my brush on the fire for light and warmth. The

Anatosaurus, a non-crested (flat-headed) hadrosaur from the Late Cretaceous of western North America. Copyright © Gregory S. Paul, 1984.

aroma of the pungent sage blended with the frying meat, the sizzling sweet potatoes . . . there is no place like home on the range!

It occurred to me I could examine my fossil while I ate. I dragged it to the fireside, where the blaze lighted the mould that had been a part of the encasing matrix of a skull. What manner of creature had I found? The teeth were small and numerous, supplemented by pegs here and there about the palate. The long and rounded snout resembled a crocodile, but a faded memory told me crocodile's teeth were all long and large. Some sort of dinosaur? I wanted to think so, but the shape didn't fit my vague conception of a dinosaur's skull. Could it have been some sort of salamander? If so, he had been a giant of his kind, and he may have known dinosaurs.

The night was growing windier and colder. I wrote up the day's events, like a real scientist, then undressed and crawled into my blankets. Through a crack in the slatting canvas, I watched the fire die. The coals burst into a last flicker of flame, as though red eyes widened to take a last look across the desert. The waver of light played across my fossil skull.

The events of the day raced through my head and drove sleep out. I ran over half-made plans for running down the west coast of Mexico to escape winter and weighed these against the desire to stay right here and scout further. When I found the rest of my strange creature, I would go right on with this work. I toyed with the thought, weighed it with ignorance and care. A dinosaur hunter . . . that's the ticket . . I would become a dinosaur hunter. I pictured myself in the field for some big museum, collecting prehistoric animals. Every single day I would have adventures like this. I might even become famous. Best of all, I would never need to work again at grubby, ordinary jobs, doing the tiresome things others are forced to do. I glossed over the fact I had no training in these matters, no schooling.

Dozing fitfully, I awoke when the moon was rising, flooding the desert with its cold light. The restless wind tugged and fumbled with the camper's canvas. A thin howl rose above the random little night sounds, barely audible before, distinct now. The howl was followed by a series of high-pitched yips, a wavering cry. And again, plaintive, mournful, an animal in pain of the moon's rising, of the cold wind's moaning, of his own lonely passing in the dead of night. I had never heard a coyote before; I knew I heard one now.

Natural mold (impression in the rock) of the palate of *Stanocephalosaurus*. This is the amphibian skull found by Bird in northern Arizona. Photo by H. S. Rice. Courtesy Department of Library Sciences, American Museum of Natural History.

2

THE SUN WAS UP before I was. The long, lean shadow of every dumpy creosote bush and sage clump told me it hadn't been up for long. Fat dewdrops on the varnished leaves of creosote and the silvery mist covering the fuzzy sage leaves told me it was cool, plumb cool.

I was almost afraid to look at the flat rock by the grey ashes of the fire, afraid to scrutinize in the cold light of dawn, afraid it might prove to be less . . . But no, the print was still there, still plain and sharp. I hurried through breakfast, as driven by the exciting prospects of the new day as by the chill air.

The climbing sun took the edge off the chill rapidly, and the air was still. There didn't seem to be as many rocks to examine in daylight as there had been in yesterday's fading twilight. However, I had covered only a few feet of the wash I had stumbled through hurriedly and fumbling half-blind the night before, when I came upon a small fragment of bone encased by sandy matrix, clearly a part of the slab I had found. Though it didn't fit, I was off to an encouraging start.

So I plowed into the task ahead, properly encouraged. By the time the sun got its temperature up and was breathing hotly on the back of my neck, I had turned over tons of rock. I climbed to the edge of the caprock and made certain no bone was showing there, no fossil. Only the large mass at the head of the wash remained as a possible source of the balance of my fossil. It weighed tons and was wholly in sight, of course, except for the detritus-covered bottom and a buried end. I searched for breaks along its front, noting that the brown surface was as weathered as old boards.

The bed of the wash below the big rock was shored up with coarse sand, a foot or more deep; perhaps what I sought was buried there. I brought out a small shovel I always carried, and went to work. At the first stab into the sand, I struck a rock. I dug it out and brushed off the surface hopefully. It was just a rock. Foot by foot, I prospected the sandy shoring along the face of the block. By hot high noon, I had to show for the day: my initial bone scrap, several blisters, and a good appetite.

As I gnawed away on a cold sandwich and gulped from a can of cold tomatoes, I puzzled what to do with my palate imprint. Leaving it behind was, of course, unthinkable. I could carry it on top of my camper, but not for long; I must ship it home by freight or express. My father would be delighted to receive it. He could have it identified at the American Museum of Natural History. But what would they say about it? I knew too well. "Why didn't your boy pick up the other half?" The first scientist to ask the question should be right where I was.

The corner beyond the end of the butte still beckoned, virgin territory. More of the ripple mark stratum, though clearly my fossil didn't come from exactly that level.

Another butte beckoned, a few hundred yards off. And beyond that . . . I cleaned up my lunch, grabbed a hand axe and a cold chisel from the van, handy tools for stone-cutting, just in case. With such luck as I had in my first hour here, I still might find me a dinosaur. Wide expanses of the ripple-mark stratum between here and the neighboring butte gave proof I was still in a comparable time zone.

At sunset I came dragging my way back to the Harley from a few buttes away, my axe and cold chisel clanking forlornly in the carrying-bag that should have been filled with fossils. Stone copies of creatures? I hadn't even turned up a piece of petrified wood. But I had noticed, from high on the flank of a butte I had climbed, another group of

Fossilized logs in a dry wash, Old U.S. 180, Petrified Forest National Park. Photo by Ken Piazza.

Its base was littered everywhere with fallen blocks from the caprock stratum. Across the breaks, the bedding showed bands of coarse grit and pebbles . . . smooth, well-rounded pebbles that must have been transported a long way in the channel of a meandering stream. The bands were never continuous but crossed and dipped like piled sand in a heavy sandbar. These ancient river sands, I thought, must have been a catch-all for anything lodging against the drifting bars . . . old logs, dead animals, floating drift and rubbish. If I didn't find something here, it must be simply because the caprock showed only a small fraction of what must lie entombed in the sweep of the mass across the top of the butte. I tried to make out some indication of the direction followed by the water that had deposited the assorted trash, but everywhere the capping had been too squarely broken.

I trudged west, prospecting. A light-colored fragment between two blocks of stone caught my eye. The inch-long chip was hard, heavy, its broken edge sharp and flint-like. It was striated, like petrified wood, but it lacked the variegated reds, yellows and blues that characterized the logs of the Rainbow Forest, resembling instead white pine but a pine half-mouldy and invaded by dry rot. One of the flat faces showed the detail of the wood's fibrous structure.

I looked up the steep slope. In the shadow of a clump of sagebrush gleamed another whitish chip and another. I stretched to gather them in, manipulated them to see if the fragments fit together. No fit. I climbed to a little ledge under the caprock and stepped on hundreds of fragments, as if a shower of them had fallen from the sky. Above me, under shadow of the caprock, was a gray mass, round as a turned fencepost and about that size. The end, broken and weathered, was of the same material as the chips at my feet. Certainly the bole of a petrified log. Possibly I was the first human who had ever seen this tree. The ones in the Petrified Forest were showier, all in their rainbow colors, but this was mine, all mine. The coarse matrix and the rolls of pebbles strung through the bedding, the rolling sweep of the bedding planes as well as the supine position of my very own fossil tree—all pointed to the force of the angry waters that had swept this particular tree to this particular point ages ago. The ranger at the park had explained these fossil trees. "The climate was different then," he said. "Wet. Forested. The trees were mostly conifers, related to the modern Arucarian pines of Australia and South America."

buttes a bit closer to Winslow, perhaps half an hour away. Not having been investigated, they seemed better prospects than the ones where I had spent the day. "I'll look them over in the morning," I promised myself.

The lonely coyote of the night before returned with new-found friends. At a goodly distance, they serenaded me through what seemed most of the night. They, you see, had been sleeping away the long daylight hours, while I had been working up a real incentive for a good night's sleep.

Next morning I piled the rock slab with its strange impression tenderly on the top of my van; after yesterday, its value had become greatly enhanced. The roughness of the road back toward Winslow threatened to bounce it off from time to time, but there wasn't room to put it inside. The rough dirt road met U.S. 66 just west of town. The pavement carried me toward the new buttes. The first butte seemed to me too close to the highway to be productive of anything worthwhile, but one to the north, more distant from the road, struck me as promising. A dim trail of old automobile tracks waddled uncertainly in the direction of the red-and-brown layer-cake of rock.

I circled the butte, disappointed to find no more fossil trees, nor signs of bone or other fossils. The next butte had been reduced to the lower sands and clays, weathered down into a few rolling, grassed-over hills. Not a favorable spot for fossils, in my newly acquired expert opinion. But a rounded grey shape on a slope a hundred yards or so away lured me on. Another petrified log. Near it, a second, almost buried in the grass and where the hill fell away, a third. Each lay exposed its entire length, where erosion had carried the hill away.

Like the trees in the Petrified Forest, the logs were partly shattered, probably jarred a good many times as they were carried down from the caprock. I pulled a piece from the largest. It came free like a chunk split for a stove. A large knot ran to the core of the tree. All the old knots and burls were as clear as they had been in life. The small splinters looked as if they might kindle at the touch of a match. If I racked up some of these chunks into a stone wood-pile just for fun, they might make a picture. After all, an embryo scientist should take a little time out for play before he buckles down.

When I had stacked up a pile of stone stove-wood that might have graced the doorway of Barney Rubble or Fred Flintstone, I finished off by rolling up a chunky chopping block that was all I could handle. Setting my camp axe into a crack in the block, I set my camera up with a delayed timer, so that I could be in the picture. As an afterthought, I placed several neat stone chips beside the axe. Then I shot myself, because a funny picture should have a little life in it, and I was all the life in sight. A silly bit of business. But after all, it was only yesterday afternoon that I had begun to become a scientist. Perhaps staid maturity would come later; only time . . . something more than a day's time . . . would tell.

When I had put the camera away, I browsed about, looking for I knew not what. Another tree, near-buried in the grass, caught my attention some yards away. It was bigger but more disintegrated from heat and cold and rain and frost and from the fact it had been somewhat in a state of decay when fossilization had set in. I kicked a piece of it loose from the exposed heart. It bounced away lightly, trailing dust like a thin smoke. Smoke, for Pete's sake, from a piece of petrified wood! Had I gone off my rocker or had Mother Nature flipped? Almost hesitantly, I followed and picked it up.

Light and porous, like wood worm-eaten or rotted to a shell, it was full of ant holes, of thin walls and winding passages through stone. It was so fresh and real I should not have been shocked

Bird and V. Theodore Schreiber with the Harley set up for camp, near Laredo, Texas.

had beady-eyed stone ants come boiling out of the stone to nip at my fingers nor surprised had they moved in and taken over these mini-catacombs.

I dropped beside the old log and abandoned myself to the enchantment of my new discovery. The little tunnels ran the log's entire length. There had been no specimens like this in the museum at the Petrified Forest. This tree had been alive with ants . . . two hundred million years ago. A feeling of remoteness came over me. A breeze wafted me from the present to the time when this tree had stood tall, long before it had fallen prey to ancient ants, when the same air had made up a breeze that sighed and soughed through the ancient pines standing shoulder to shoulder in the forest on the flank of an ancient shore.

The salamander-like monster whose partial stone remains lay by my machine must have slithered through the mud and water here, may have gazed with dimly puzzled eyes on tall creatures striding along the shore on two legs . . . the dinosaurs.

I ran my fingers along the log. Perhaps it had been beetles instead. Perhaps worms; I didn't really know enough to be sure, yet. But of one thing I was sure: both the museum at the Petrified Forest and my father back in Rye, New York, must each have a generous sample of these interesting fossil holes. But for my father, a rabid entomologist, I might never have stumbled even this far into the shoreland of science. From the time I was big enough to scurry along beside him, he had taken me along on innumerable field trips in searching out the life cycles of insects. Though moths and butterflies were his special passion, he was forever prying into every fallen tree in the woods around our home. "I like to see what lives under the bark," he would say.

The breeze stiffened. Conscious of a great emptiness, I sighed like an old Arucarian pine tree of long ago and looked at my watch. Emptiness? No wonder; it was midafternoon and I had forgotten all about lunch. Returning to the van and the Harley, wolfing a sandwich, and packing up for the road were the work of a few minutes. I strapped the fossil jawbone to the top of the van to be shipped home to Father, dropped the samples of the fossil insect borings into odd corners, and took to the road again. Winter was coming on in Arizona, and I had more or less decided to dodge it by going south, to return here, perhaps in better times.

Someone along the way had told me I should try wintering in Mexico, in a place called Mazatlan. They had a lot of poor roads there, I'd heard, grading all the way down to none and all the way up to oxtrails, but motorcyclists always claim they can go anywhere a horse can . . . and never need worry about corn and hay.

I left with a sneaky suspicion I'd see this place again.

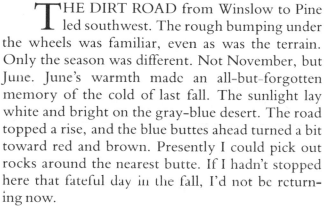

3

THE DIRT ROAD from Winslow to Pine led southwest. The rough bumping under the wheels was familiar, even as was the terrain. Only the season was different. Not November, but June. June's warmth made an all-but-forgotten memory of the cold of last fall. The sunlight lay white and bright on the gray-blue desert. The road topped a rise, and the blue buttes ahead turned a bit toward red and brown. Presently I could pick out rocks around the nearest butte. If I hadn't stopped here that fateful day in the fall, I'd not be returning now.

Why had I come back? It had seemed simple yesterday, still a day's ride to the east. Simpler still in Laredo, Texas, several days before. But now, the butte was here, and I was here, and my optimism amazed me. Certainly I had no sane reason to expect to find the missing half of last November's fossil skull.

This return trip had begun on a park bench outside the post office in Mazatlan, Mexico, where I read a letter from Father.

"The fossil skull has proven a real discovery," he had written. "I had it identified by Barnum Brown, the dinosaur and fossil reptile man at the American Museum. The rock was too heavy to carry around New York, so I made a life-size outline sketch and took it with me. You should have seen Dr. Brown's eyes light up when he saw it! He immediately pronounced it a new genus and species of stegocephalian, an amphibian somewhat like a giant salamander. Then almost in the same breath he asked, 'Where's the other half?'"

The other half? I squirmed uneasily on the park bench and read on. The curator of fossil reptiles was planning a description of this hitherto unknown creature, but it would be incomplete unless the other half of the skull cast could be found. What had been an interesting discovery by a curious boy had turned into a matter of at least minor moment to science. Hot dog! And now, what could I do about it?

So here I was, doing about it. I pulled up at the foot of the little wash. All signs of the first search, the tons of turned-over stones, the hours of futile digging had been erased by the winter's rains and melting snows. I shrank from the apparently hopeless task of turning the rocks and sands of the wash over a second time.

Half a mile away, a few Herefords grazed on an open flat. They reminded me of an additional prospect I had thought of yesterday. Roving stockmen and cowboys sometimes pick up unique rocks from land they cover on their daily rounds. Perhaps the missing half was lying in the yard of the rancher who ran cattle here. I should have stopped when I passed through town and inquired for such a person.

I plodded to the spot where the palate imprint had lain last November. I looked up at the caprock from which it had crumbled away, then right and left at the tons of tumbled rubble that sloped all the way down to my feet. Tons and tons and tons! From far-off Mazatlan, inspired by Father's letter, it had all looked so easy. And of course I had looked to winter's erosion forces to have already handled the hard work. And, in truth, winter had moved a lot of stone.

Wearily I trudged back to the camper, dug out my pick and shovel, stumbled back to the job, already fatigued by the utterly hopeless prospect. Deciding to begin just above last November's lucky spot, I planted my weary feet a step or two up the talus slope from the point of last November's lucky find and looked up at the great banded block that had spewed it out. Hopeless.

I kicked idly at a hefty rock at my feet, disgusted to have crossed half a country on such a wild goose chase. The rock didn't budge. Nor did it release an avalanche of the tons of piled-up rock that awaited my attention. Disgusted, I knelt and struggled mightily to roll it over.

Then I sat down on a rock conveniently located behind me. Luck never runs this way! There, right there at my feet, was the other half of last November's find. "Where did you come from?" I gasped. No answer.

I looked up the wash and got an answer. Water cascading down the wash from the winter's rains had uncovered the end of the only rock I had passed over; I remembered it well. It had been partly buried, too big to roll, too buried to dig out. And it hadn't looked promising, no signs of breaks or cleavage lines. But winter had broken it neatly in half and dug it out of the sand.

Almost reverently, I knelt before my free prize. The skull itself was missing, possibly fallen away and disintegrated, but the retained impression was as sharp and clear as the palate imprint. Here was the happy end of a hopeless quest. Sudden. Unexpected. Now what?

I tripped lightly back to the motorcycle van to dig out a cold chisel and my all-purpose hatchet. Somehow I must release the specimen from the still massive rock of which it was a part, first cutting a U-shaped channel around the impression. As soon . . . and not so soon, at that . . . as I had channelled a bit deeper than the brown cast, I must split it off with a lateral cut beneath the face. I tested the surface with my chisel. Only brownish-red sandstone, after all. But like well-set concrete. A good bit harder than I had expected. And this had to be a perfect job; I had to pay off Lady Luck for this fantastic break; it would be a sin to crack this specimen.

At first, time passed at a fair rate. I could already visualize the brown outline of skull on a neat stone block resting in the Museum with upper and lower palate; I could see Father and Barnum Brown enthusing over it and planning a name for it. I didn't know much about Barnum Brown, only that he was one of the three Kings of Dinosauria, and the only one of the three still living. Only that he had dug up and assembled more dinosaurs than any man living or dead. Only that now he could complete his paper on my specimen. Only that Lady

Luck had forged a link . . . however tenuous . . . between us.

Noon came and went unnoticed. Daydreaming tided me over skinned knuckles, rendered me unconscious of the fact this was some of the toughest sandstone ever forged in Nature's foundry. By evening I was arm-weary, knuckle-skinned, daydreamed out.

Next morning bore out my findings of the day before. The skull imprint was no softer. By lunchtime I faced the facts: I must develop a speedier technique or chisel rock all week. My channelling was deep enough, but undercutting promised to be an endless task. Could I free the cast by drilling lateral holes below its level and splitting it off by use of rock wedges? I had no proper tools, and only a vague idea of how this is done in quarrying.

In Winslow I purchased a long rock drill of suitably small diameter and a number of heavy iron spikes. Back at the wash, I heated the spikes in a sagebrush fire and flattened them into wedges. By late afternoon, I had drilled three holes laterally under my precious cast and felt sure I was on the way to success.

Next morning, after extending the lateral holes just a bit, I put in my wedges and started tapping them gently with the head of my hatchet. There were a few nervous minutes as I drove in the wedges and increased the stress. Then crack lines began to show between the wedges, and the specimen dropped free to the sand below. I was jubilant.

I moved the precious burden . . . that's right, burden . . . to the top of the van and went back to examine the area. When I finished, the place had been covered with a fine-tooth comb and I was sure the rest of the creature's skeleton had been, like the skull itself, broken up and scattered by the same forces of nature that had revealed the palate imprint and had buried or destroyed the skull itself. Finally confident there were no more remains of this or any other creature in the vicinity, I mounted my trusty Harley and set out for U.S. 66.

It was good to be back in northern Arizona. The map told me the Grand Canyon lay to the north and west. I would have visited it last year but for the onslaught of winter. There were also other colorful expanses of the Painted Desert still to see, over around Cameron. I intended to see both but couldn't make up my mind which first.

The Harley's wheels rolled me past the low hills where I had found the ant-riddled petrified wood, out of the last of the buttes, and onto flat desert. The last faint tinges of red disappeared from the desert's sandy clay. The trailing skirts of the formation the geology people called Queen Triassic swung far off to the west, north of the valley of the Little Colorado. The road began to dip toward a canyon, a rough gorge cut deeply into a dark grey stone. There was a parking place on the canyon's rim, and a sign reading "Canyon Diablo." I pulled off the highway to study the rock of the canyon face. No great shucks, as rock. A dull, dark, weathered face. But here and there, in protected cracks, crevices and crannies, a limey white. A low wall erected for a short distance at the canyon's rim was made of this same rock. Its freshly quarried surfaces showed it to be a shell-laden limestone, compacted from shell-filled mud once accumulated on an ancient ocean floor.

I knew from charts and maps back at the park center that this was under the Triassic formations, the dawn-time of dinosaurs when the great creatures were just beginning to take over the world. The fossil shells were not such modern types as clams and oysters but older, unfamiliar species. I knew an inland sea had covered this part of Arizona before Triassic time, but what this formation was called I had no idea. Sorry I hadn't studied the park center charts more, I realized I had some to go; a young fellow with such a tenuous link to Barnum Brown should improve his mind all he could.

The road wound and wiggled west toward Flagstaff. The desert swept slowly up toward the dark slopes of the San Francisco Mountains. Presently the sparse desert flora gave way to low piñon pine and scrubby cedar; in a few miles taller sugar pines thickened into true forest. Stuck in a hole in this forest was Flagstaff.

I hunted out a lumber yard, where I bought wood and nails to package my prize and got permission to build a box in the rear of the yard. An oldish man in overalls brought me my boards and started slightly when he saw the rock on top of the van. He ran a roughened finger along the outline of the long snout of my fossil.

"What," he mumbled around his 'chaw' of tobacco, "would you call that dern thing?"

The man was a long way and a far cry from Barnum Brown, but I shouldn't neglect any chance to practice showing off. "That happens to be the stone cast of the top of a stegocephalian skull," I explained. "It's about two hundred million years old. The entire creature, alive, would have looked a lot like a giant salamander."

The old yard clerk peered up at me through shaggy brows. He scratched his head, gave it a quarter-turn, and ejected a copious mouthful of tobacco juice. It was not an expectoration of contempt; it was more like an ordinary mortal saying, "Wow!" or giving out with a long whistle. I accepted it as a tribute, even before he added, "Well! I'll be jiggered!"

I turned back to my task, and the old man pattered off. But presently I heard footsteps. Three men, the whole office force, were approaching.

One of the men said, "Old Dan tells us you've found a rock with some animal's head printed on it. Could we have a look? Is this it?"

Again I explained the origin of the specimen. They listened, quite obviously enchanted. The first to speak said, "Well, I've heard of such things being found in Arizona, but I never saw one before. Now you take petrified wood, of course . . ."

"You know who ought to see this?" another interrupted. "Rad Linderman, across the street. Why don't one of us call Rad?"

The three men turned back to the office, and quiet descended on the yard again, except for my sawing and hammering. As the packing box took form, a tall, sandy-haired young man came around the corner of a shed. He wore the uniform of a filling station attendant, but his eyes were shining like Father's when he had just come upon a rare insect.

Breathing hard, he said, "One of the boys out front told me about your fossil. Looks like you've made quite a find. Where are you sending it? Back East to some museum?"

Here was a kindred spirit. I sat down on my embryo packing case and went over all the details of my find, at Rad's request. He listened well. Then we went into his interests. He collected Indian things as a hobby and told stories of old ruins near Flagstaff, where he had done some digging. Then the talk shifted back to fossils.

"There's a place on the Painted Desert," Rad said, "the other side of Cameron, where there are more of the same kind of trees you saw around Winslow. Then, not far from Tuba City . . . and at Cameron, too . . . there are dinosaur tracks."

"Dinosaur tracks?" The box-making was forgotten.

"Yup! But I've only seen the ones at Tuba City; the ones at Cameron are harder to get to."

A dozen questions crowded into my mind, but Rad rose suddenly from his keg. "By Gosh, I've got to get back on the job or hunt a new one." He shook my hand and prepared to run. I caught his sleeve.

"How do you find these tracks?"

"The ones at Tuba City aren't far from the road. But if you want to see the ones at Cameron, I think the trader Hubert Richardson could tell you. You look up Richardson when you get there."

Then he was gone. I completed my box-making and packing and hunted down the express office. Just before supper time I wandered into a hotel lobby in search of a newspaper. A number of travel folders were on display in a rack on the wall. One bore the word 'Cameron' on the colored cover and advertised the delights of living in a small, pic-turesque hotel, owned and operated by the Indian trader Hubert Richardson. I turned the pages, look-ing at the illustrations. One was a photograph of a large dinosaur track. It was three-toed, impressed deeply in the solid rock. Another picture showed a few men clearing a flat exposure of sandstone which showed several such tracks. A name leaped from the caption.

The man to whom I had just shipped the fossil skull was in the picture. It was Barnum Brown, ac-cording to the story, who had uncovered the tracks.

The Grand Canyon had just lost its chance at me. I knew where I'd head in the morning. The footprints, the folder explained, were at the end of a trail leading to Dinosaur Canyon, sixteen miles east of Cameron, on the desert.

Dinosaurs from the Latest Cretaceous of western North America. The huge carnosaur *Tyrannosaurus* attacks an adult and juvenile of the gigantic ceratopsian *Triceratops* in this lively restoration. Copyright © Gregory S. Paul, 1984.

4

THE PINE FOREST thinned out as the road north from Flagstaff left the slopes of the San Francisco Mountains. Tall sugar pines gave way to piñon pine, creosote and sage, and finally to scrub desert where only the horizon bounded vision.

The road dipped slowly down the gentle slope that was the west flank of the valley of the Little Colorado. Across the valley, faintly discernible below a row of distant cliffs, lay a band of many colors. The reds, blues, mauves, maroons, the browns and greys of those hillocks and buttes that had marched westward from Holbrook and the Petrified Forest turned at Winslow to follow the east bank, losing themselves with it in its northwest course. They were the edges of that same great sweep of Triassic rock and clay south of Winslow from which those other scattered buttes were carved. They were swinging along with the road, hugging the east bank of the Little Colorado, and would meet it at a point below a series of jutting benches just ahead.

The giant steel piers of a great suspension bridge with hanging webs of black cables confronted me suddenly. The road dipped over a red rock bench and brought into view a small hotel, built of the same red sandstone. Scattered tourist cabins and a large store were grouped on the barren rock near the end of the bridge. The store, the Indian trading post of Cameron, was dominated by a large false front decorated with two rows of three-toed tracks, replicas of dinosaur trails, done in black paint. The road led into a parking place between the store and its gas pumps. Seated on a bench near the door a heavy man sat conversing with a copper-colored Navaho with long, black hair. This could only be the trader, Richardson. I inquired about the trail to Dinosaur Canyon.

"The trail to Dinosaur Canyon?" The trader looked up and repeated my words. "Why, boy, there's no trail to Dinosaur Canyon."

He read the disappointment in my face and stance and looked east to where the desert lost itself at the base of the distant cliffs. "What's left of a trail takes off at the other end of the bridge," he said. "But it's washed out in some places and drifted over in other places with sand. I don't suppose anyone's been over it in two-three years. You'd lose it before you went three miles."

He turned back and resumed his conversation with the Indian. I listened to the measured grunts and sing-song vowels of Navaho for a minute, concluded I had been tuned out, and wandered away, dejected. But the trail had been lost only since Barnum Brown had been here. And I was freshly back from five months on Mexican roads; my old machine and I had found our way through quite a few places we were told were impassable. I rolled the machine over to the water faucet by the gas pumps and filled every water can I owned. It was still early in the day, and Dinosaur Canyon, after all, was only sixteen miles off.

The red walls below the big bridge were sheer as the sides of a box. Solid, unbroken sandstone, thick. Through the gorge ran the dribble called the Little Colorado River. On the other side, the trail turned east for a short distance. It followed along the rim of the canyon over a bare expanse of red rock, then turned in the direction of a line of gradually ascending ridges, still eastward. The ridges were the lower portions of the colorful Triassic formation that rolled and dipped toward the still distant cliffs.

The trail climbed slowly and crossed a low gap in the first ridge. Here sand blocked the way, a long dune of wind-drifted sand, too wide and deep to cross with a wheeled vehicle; the machine would bog down in no distance at all. The cliffs seemed as distant as ever, but I could see features now not visible from Cameron. The lower part of the red and weathered ramparts broke down into what seemed an advancing army of little stone gnomes. Those close in to the face were in close ranks and stood tall, but out on the desert they were scattered and decreased in height, like corn rows where the rows end in the shade of the trees at the edge of the field. It wasn't shade that affected the size of these curious formations; the sun glared down on all alike and made of them a picture so strange I was impatient to cross this sand barrier and be off between these ridges where the wheels might roll in among them. There I should surely find the dinosaur tracks!

A red clay outcrop extended southward from where I stood, trackless, but fairly firm. I followed it south until a break in the paralleling sand dune let me cut east to cross the next ridge. Beyond were other ridges, but for every yard gained east toward the cliffs I had to travel south four or five. I finally reached a point where the ridges had been cut through, as if by a great knife. A wide dry wash lay below, away below, coming in from the east and departing in a great sweep toward the canyon of the Little Colorado. Its yellow, water-swept floor looked hard-packed. If I could find a shelving bench to lead me down to it . . .

A hundred feet farther on I got my break. A bench sloped gently back with the dip of the strata, forming a descending ramp. I could get down, maybe. Could I climb back out? Maybe.

The long incline at first furnished ample passageway, but narrowed, leaving scant room to squeeze between the wall and the edge. The wheels clawed for a footing, the sidecar clinging precariously to the brink. Bits of rock rattled down into the canyon below. I inched along, holding breath and brakes. For an instant the crankcase hung up on a boulder, scraped free. I breathed here. Then I was down in the wash, and in a moment swinging merrily upstream at thirty miles an hour. My breathing improved.

The hard sand made fine going. The wash twisted and turned, as little riverbeds have a way of doing. I turned a bit to dodge a rock, and the wheels hit a patch of dry sand that pulled at the tires as if lassoing them. The machine's momentum carried it to firmer going, but the incident was a warning not to cross such places at a low speed. A

fork in the wash appeared. I went left only to keep rolling in high gear.

The old Harley took the good and bad of the wash in a wild dash; a mile clicked up on the speedometer. Another. The walls of the canyon, multicolored curtains moving by like a motion picture, closed in on me gradually on either side, decreased in height. Low shelves of rock appeared across the path ahead. I shifted into low gear, tackled the last low bench, which went sharply upward. The sidecar wheel caught for an instant, pulled free, and we topped the last rise with rear wheel spinning. I was out of the wash, but barred in all directions by rough going. The smell of hot rubber and smoke from an overheated engine told me something. The foot wall of a great cliff a quarter-mile away told me more.

But close at hand was the first of the gnome-like stone figures I had seen away back. Hardly gnome-like now, it loomed as high as a house and leaned crazily to one side, a falling mass arrested in mid-air. From one angle it looked like a pile of stone biscuits; from another, a heap of racked-up hats. On my other side was a similarly distorted column, neither standing nor falling, defying gravity and reason. Beyond, another, and another, ranked around in all directions. Had anyone told me about these, I doubt I'd have believed them.

There were no blended rainbow colors here; the basic color of the rock was red, but red as lipstick. Across this gaudy base was streaked a ghastly pallid blue. The hodge-podge lacked pattern even as the shades lacked harmony. It was as though Nature, after carefully applying the red, had flung the pot of blue at the rock in frenzy. The result created faces on the knobby biscuits. Long smears became slanted mouths; rounded blobs were staring eyes or misshapen ears; occasionally circling bands added here a scar, there a wrinkle. The midday shadows, black and narrow, hung below protruding chins like false beards.

What had shaped these strange formations? Undoubtedly they had once been part of the cliff, but as the wall had melted back and wind and rain had carried away the dislodged sand, these masses had remained behind. I stood gaping while these rocks made faces and rude remarks in stone in a way I could sense but not hear. Suddenly I recalled my mission: where had Barnum Brown found the dinosaur tracks? Where, indeed, was Dinosaur Canyon?

I dug in the van for the travel folder containing

the footprint picture. In the foreground lay the flat expanse of stone marked with the three-toed tracks; in the background, against a cliff, a projecting rock bench. I studied the cliff in front of me, rising a thousand feet or more. Some distance south, toward the lower end of a great bay, I made out what was surely the jutting bench shown in the picture. All I had to do was place myself at that point, on foot; the roughness of the area put a stop to further driving. Apparently the area cradled in the great bay, rather than the narrow defile, was Dinosaur Canyon.

Judging the site to be no more than an hour's walk away, I belted on a canteen of water, slapped together a few sandwiches, grabbed a Kodak, and took off. Over bare, baking rock. Under bare broiling sun. And they say it was a warmer world when the dinosaurs came through here! The rude stone gnomes thinned out, disappeared, and I was left all alone. The far side of the great bay, the bench I trudged toward, remained as far away as when I started out; clear desert air makes distances deceptive. A lone stone tower on my right wavered and shifted with the immediate landscape, and I moved, and time moved, but everything else stood still. "This place is hoodooed," I said. I got no answer.

Ahead and a bit to the left lay an isolated group of the grotesque stone gnomes, varied and variegated, tending more to height, slender pillars topped with ungainly big heads somewhat like mushrooms, making faces at me. One more spectacular than the rest made faces at me, and I got an urge to detour to the left a bit and push his fat head off his skinny neck. When I got there he had grown a lot bigger than I was and he had changed his appearance; he was a fat Humpty-Dumpty with a satisfied grin on his face, staring out over the desert without a worry in the world, not even seeing me, oblivious. The next big rain and strong wind, the tremor of a thunder storm, might bring him down. I took his picture and almost tiptoed away.

So it came to pass that I finally stood at the foot of the cliff, and the cliff stood, too, hanging in the sky above me, an unclimbable mass and maze of eroded rock. Here below, on either side, spread a whole square mile of flatland that might anywhere hold Barnum Brown's dinosaur track site. I wandered east, north, southwest, east again, a great crisscross of the area. No sign of a long, flat, exposed surface bearing three-toed footprints. No sign of a road, nor tire tracks, nor evidence of invasion by humans. No restaurant. So I ate my sandwiches. With sand and gusto.

What landmark, other than the bench on the

Capstone rock atop a bentonite hill, Old U.S. 180, Petrified Forest National Park. Photo by Ken Piazza.

cliff, might aid me now? One of the grotesque rock formations had stood out in the foreground on the travel folder, but which one? The folder was carefully stowed away in my camper for future reference. I plodded randomly through the inferno, scanning countenances of innumerable stone devils in search of a familiar face. Oh, for that travel folder!

I rested awhile in the friendly shade of one of the stone gnomes, an old fellow who looked a bit senile but who had not made rude faces at me. Kindly, silent, complacent, he may have quietly induced me to give up. Too bad about the tracks, but it was high time to start back toward the Harley. Tramp, tramp, tramp. Slowly the cliff fell behind, very slowly. Again the illusion of trudging on a treadmill, of walking nowhere at no speed . . . a depressing optical effect of clear desert air.

Crossing a long stretch of open sand, I came upon a small dry wash. It was no more than two or three feet deep, twenty or thirty wide, but at its bottom lay a patch of familiar dark red ripple-marked sandstone. No longer was I in a desert inferno; I stood on the shore of a Triassic sea. But I was still all alone at the end of the ripple-marked patch. Just ahead was another, not very wide, but with a series of strange markings. I brushed away the sand. An unmistakable three-toed track! And another, and another, and another! Not widely splayed, the toes pointed straight forward, and where the knobby end had pressed was the deep scratch of a sizable claw.

I forgot the place, the time of day, the importance of getting back to the camper, and trailed a dinosaur. The tracks crossed the wash diagonally, until the trail disappeared under the rock wall. I longed for a crowbar, to trail him to his lair. I had a pocket knife. His stride was about three feet, which would have made him just better than my size. The sharp claws indicated a carnivore. A little fellow, but big enough; perhaps it was best we hadn't crossed here at the same time.

A shadow creeping across the brick-red stone caused me to look up; the sun was about ready to call it a day behind the cliff; if I didn't head for camp now I'd never find it tonight. I couldn't face up to a night among rude stone gnomes with carnivorous dinosaur tracks all about. I climbed the rim of the wash and hurried for home. It was trudgy drudgery. Night breezes crept out of the crannies of the great stone wall, but the rocks underfoot still radiated heat like stove lids. The stars came out, glit-tered brightly, and made of the stone gnomes fresh new monsters crouching in the dark. I glittered less brightly; I walked a good piece beyond my machine, missing it in the dark, and didn't find it until near midnight. Without cooking, I wolfed down what came to hand, gulped thirstily from one of my gallon cans, and turned in without building a fire.

The new day put me in a new exploratory mood. I wanted to go back to Dinosaur Canyon, but a great rock spire set out on the desert in the opposite direction intrigued me. It was visible for miles, reminding me much of Inscription Rock in New Mexico. This famous rock had lured many early explorers, travelers and emigrants out of their way, and a great many had scratched their names on it as a record of their passing. I had a notion to take a chisel with me, but didn't; I hoped to visit this rock and still to make my way back to yesterday's dinosaur trail to find for certain the tracks Barnum Brown had visited.

The great rock was unique. Not round, not square, it had something of the nature of a totem pole. Across the highest block Nature had etched a monstrous face with one eye, a drooping cheek, a twisted mouth. It leered at my approach, seeming to call me a fool to have left the land of living men for this stark place of dinosaurs and stone devils. A few massive rocks, once part of the tower, littered the base, some standing as high as my head. I moved in among them, seeking a way to the base of the tower. The stone face, its leer changed to a sly smirk by shifting shadow, was thirty-five or forty feet above me. The wall in front of me was sheer and smooth, and again I had that old common urge to leave my mark here; why hadn't I brought that chisel?

Somebody had brought a chisel. There was already an inscription carved in the stone, just a bit overhead, weathered a little, but still clear. It read:

B. Brown
Am. Museum of Natural History
New York City

I blinked and read it again. It couldn't be, but it was. Even as I had come here, so had Barnum Brown. And he had brought his chisel. I relished the picture of the Curator of Fossil Reptiles, the renowned paleontologist, standing beside this remote rock, carving his name. I felt I knew my unknown friend much better now.

5

NOT ONE BUT DOZENS of dinosaur trails crossed the face of the stone. The creatures had come and gone in all directions. All left three-toed prints with clear claw marks. Here and there were little furrows between the footprints, tail-drag marks, but oddly rare. I wondered if they had come by singly or in groups. No way to determine this now, but I assumed they had probably passed within hours of each other, while the mud was soft.

An old broom lay beside one pile of rocks, undoubtedly left behind by Brown and his assistants. I must have wandered yesterday within a hundred yards or so of this broom but hadn't caught sight of it. The strings that bound the broomstraws were rotted away, the wires holding them to the handle were loose and rusty, but it held together enough for me to sweep away the thin layer of sand and dust that partly filled some of the tracks. I was happy to serve, thrilled to think I was a belated and unofficial part of the departed expedition.

A thin layer of rock had been stripped away from an area some seventy-five feet long by thirty feet wide. Here three hundred footprints were in view . . . I counted them. All were carnivores, ranging in size from three to fourteen inches, possibly four different species. I studied them so thoroughly I felt I knew each and every footprint by heart.

Next morning I had to face up to a hard choice: learn to live without water or return to civilization. The return trip, made without mishap, was tedious rather than adventuresome. Back at the Cameron trading post, I decided to retrace a bit more and pick up mail at Flagstaff. The most interesting piece was a letter from Father bearing three postmarks; it had followed me west from Laredo, Texas. Included was a clipping headed, "Dr. Barnum Brown Off To Dig Up Two Dinosaurs." It described an expedition, just leaving for Greybull, Wyoming, headed by Brown. The bones of two large sauropod dinosaurs found by the eminent paleontologist the previous year, the article stated, were to be collected for the American Museum.

I wondered what it would be like to dig up dinosaurs. Would the bones be encased in rock, as my odd amphibian skull had been? Or would they be in clay? It occurred to me that Brown ought to be at work right now, unless he might already have finished. The clipping was more than a month old. How long does it take for a big museum to dig up a big dinosaur?

I read on, looking for an answer. The clipping was vague on this point, but a new and surprising feature about dinosaurs was mentioned by the reporter in describing his visit to Brown's office:

"Dr. Brown stepped to a cabinet," he wrote, "and brought out a tray filled with well-rounded stones, about the size of hen's eggs, and well-polished. 'These are gizzard stones,' he said. 'Some of the dinosaurs customarily picked up rocks, as chickens pick up gravel, to grind their food.'"

Dinosaur gizzard stones. From away out here, I could see the polished stones from the great gizzards of the dinosaurs the Curator of Fossil Reptiles had found. I had heard some odd things in my life, but this was the oddest. I reread that part of the article, recreating the picture and establishing it in my mind. "Dr. Brown expects to add to his dinosaur gizzard stone collection this year," the article concluded.

I knew my next move was to cancel my long-deferred trip to Grand Canyon. "I'm off for Wyoming," I told myself, "to see if I can find Brown digging up his dinosaurs." I got no rebuttal.

The way led north through Cameron toward southern Utah. I was able, with my sparse and newly dug-up bits of knowledge, to trace the Triassic beds, laid down in dinosaur dawn times, from Cameron to Marble Canyon; beyond there I lost them under a maze of cliffs. In southern Utah the road began to climb seriously, crossing other formations that overlaid the colorful sands and clays by thousands of feet. Once my attention would have been fixed on the spectacular aspects of the scenery; now I saw everything with new eyes and every cliff wall took on new significance.

In Wyoming the last miles to Greybull stretched out forever, across the flat irrigated lands of Big Horn Basin with the Big Horn Mountains in the east and the Rockies to the west. A long curve swinging over a rise gave me my first view of the town. Lying in a bend of the river, among cottonwoods, it boasted no tall buildings, but the stacks of a small oil refinery rose above the trees and dwellings of the main thoroughfare. I drove into the business district, peering into the faces of people along the street, hoping against hope to catch sight of Dr. Brown. I stopped at a barber shop on the street, to question the only barber on duty.

"Dr. Barnum Brown? The man who's been hunting dinosaurs? Why, he left town just a day or two ago. Believe he went up to Billings, or somewhere, in Montana."

My heart sank. I had arrived too late. "Did he get his two dinosaurs?" I inquired.

"Well, I couldn't say. He worked up around Shell, I understand. For several days."

The barber looked me over curiously; I must have been the picture of dejection. I might at least visit Shell and look over the site. The barber read my thoughts. "Tell you what," he said. "If you want to see where Brown worked, why don't you drive up to Shell and look up Bill Paton? Bill runs a dude ranch at the mouth of Shell Canyon. He can probably tell you all you want to know about the whole thing."

Following the barber's directions, I turned east from the heart of town, rattled across the loose boards of the bridge over the Big Horn River, and headed out a gravel road toward the foot of the Big Horns. The road paralleled a crystal-clear stream which emerged from a canyon in the distance and flowed gently down the sloping plain. Cottonwoods lined its banks, and green alfalfa fields by the wayside told of the benefits brought by this mountain stream. Beyond the point where gravity carried the waters in an irrigation ditch, open desert lay on either side. Here appeared a line of butte-

like hills, bare and sere and scarred with greedy fingers of erosion. I wondered if Brown had done any bone-prospecting among their dark grey shapes.

The towering Big Horns loomed closer. Grey limestone arched upward in a long roll as though a great giant lay beneath the elevated sheet, asleep. The lower flanks were like blankets which no longer covered the slumbering form; they had fallen down in a rumpled heap. A great red blanket, thin and ragged, its flaming edge visible for miles, rested directly against the limestone. Next came thin blankets of greenish-grey and brown, and another as bright with color as Joseph's coat. Showing from under the edge of these was the somber grey blanket that covered the broad plain just east of Greybull.

Until I passed the little town of Shell, I had been driving for miles into a wide rent cut by the creek into the sleeping giant's colorful blankets. Greenish shale or clay bordered the wayside for some distance, then red rock blazed in its place, rising sharply against the grey limestone. The alfalfa field narrowed, squeezing in under the red walls with the cottonwoods, close to the rushing creek. A lane, crossing a little bridge to the left, led to a lodge-like dwelling with a huge stone chimney at one end. Paton Ranch. I parked the panting Harley and climbed the steps of a wide porch.

A knock brought prompt response; the door opened, and a young woman greeted me. Her eyes sparkled at mention of Brown's name.

"Dr. Barnum Brown? Why, he's been working up at Howe Ranch. But from last accounts, he's left this part of the country." She looked at me with lively interest. "Do you know Dr. Brown? Are you from New York too?"

For a moment I entertained the thought I had been mistaken for a lost or strayed member of the expedition. I mumbled unintelligibly, and the young lady threw the door wide.

"I'm Mrs. Paton," she said, smiling. "Do come in and tell me all about it."

My end of the story was soon told. Brown's visit in the vicinity, I learned, had aroused considerable curiosity among the people around Shell, but the Howe Ranch was some distance north and the expedition's stay there had been brief. Little was known of what had been accomplished. The Patons had been intending to drive up to see the diggings, but hadn't yet taken time to do so.

"Bill's gone to town, but he'll be back shortly," my hostess told me. "Maybe he'll have picked up some fresh news. You must wait to see him." Then she asked, "When did you have lunch? How about some cold chicken from the icebox?"

It was midafternoon, and I had forgotten to grab a bite in Greybull. Dissenting politely but without weight of conviction, I soon found myself seated in a breakfast nook, feeding a ravenous appetite with the Patons' excellent food. Between bites, I learned both the Patons were from the East. Mrs. Paton was a New Yorker, and the American Museum of Natural History held for both of us the flavor of home. We chatted along until I felt I had used up the courtesy a random guest is entitled to, especially in the West. As I was to take my leave, the master of the ranch drove in. I was introduced to a slender man wearing high-heeled boots and a wide-brimmed Stetson. Back in the house, supper appeared more or less magically, and there was a plate for me. "It's soon coming on night," Bill Paton said. "You might as well wait until morning now, and Rowena and I will drive you up to Howe's in the car." I protested, but not enough to do myself out of a pleasant evening with new-found friends.

Before we were halfway to the Howe Ranch, I knew why the Patons hadn't driven up before. As dirt roads go, the road was good . . . for a fair distance. But as it crept up under the flanks of the Big Horns, it became rough and obviously little-travelled. In a little while, it became clear why it was little-travelled. Bill started staying in second gear. The road snaked its way about, around promontories, with vision sometimes limited to a few feet. Water-worn ruts added more twists, giving tires a fearsome punishment. I shuddered to think what must happen if we met a car coming down. A final heave and grunt in low gear took us up over the crest to safe going along a ridge. I resumed breathing regularly and risked looking around.

Ahead the road dipped into the upper end of a little valley. It apparently ended where the valley narrowed in a series of gullies and washes, beneath a split like a great keyhole in the mountain's flanks. A small ranch house, with three or four cottonwoods in front, stood on the east slope near the end of the road. There was a sheep corral below the house, and a narrow field of alfalfa lay across a gully under a ridge beyond more cottonwoods. A patch of green above the house marked a spring and the origin of the little valley's water.

"I think I see old man Howe out in the corral," Bill said. A stooped figure moved from under the shed and started toward the house.

The road leveled, and we came to a gate. Presently we came under the shade of the big cottonwoods, and Bill hailed the old man. "Howdy, Mr. Howe."

The old man stepped to the side of the car.

"We've come to look around where Dr. Brown has been at work. This is R. T. Bird from New York. He knows Dr. Brown, so it will be all right."

Mr. Howe looked me over closely. I did as much for him. Here was a man who had seen dinosaurs dug up right in his own back yard. He was short, desert-withered but with a shrewd and lively twinkle in his eyes.

"Well," he said, "all I can do is show you where the bones are buried."

Mrs. Paton inquired after Mrs. Howe and the family. The old man assured us all was well with the folks and began to talk of this year's crops. Conversation shifted to the sorry price of wool and rambled on and on. I got out of the car and stretched.

"The bones are across that hill," the old man said, "the other side of the corral."

He set out for the corral gate. Bill Paton and his wife got out to walk along with us. The gate was a heavy ten-foot affair, sagging on its hinges, leaning toward the opposite post, against which it swung. It was held closed with a heavy rock propped against it. Mr. Howe stooped and heaved the rock to one side. It was curiously smooth, and rounded on the bottom like the knuckle end of a bone.

It was a bone. I stared, transfixed. The end of a giant limb in stone was used to prop a sagging gate. Mr. Howe turned it on the knuckle and let it fall. It thumped heavily on the hard-packed dirt, broken end up, and dust rose from where it dropped. I knelt beside it, took it in both hands and rolled it up on the rounded end. About a hundred pounds, dark grey, edged with black, hard as flint. I turned to the old man.

"Good Lord, where did this come from?" I asked.

"Oh, that's jist one o' them," he chuckled. "Lots more over on the hillside, like I said. I call 'em 'my old bones.' They've been layin' here ever sinst I can remember."

We crossed a gully and reached the hillside, low and overgrown with sagebrush and tumbleweed at the base, but the grey clay slope above was fairly open and clean. I saw a piece of a huge rib lying in the grass. I picked up the dark fragment and realized it was but one of many. The Patons were picking up and examining other fragments.

"Brown gathered up the biggest and best of 'em," the old man said. "Buried 'em in a pile at the top."

We followed the old man up the hill, along a recently beaten path. It led to a flat area about the size and shape of a barn floor, stripped clean of its brown sandstone cover. The old man walked to the east face, where piles of fresh dirt lay. "What bone Brown uncovered, he uncovered here . . . and covered up again," Howe said.

I looked about, over the area that had been stripped. "Why did he move all this dirt?" I asked.

"Oh, there's more bones, right where you're standin'. I guess he thought they was too many to take up this year."

"Oh, so he's coming back? Next summer?"

"Figgers on it."

The Patons and I walked back and forth across the top of the little hill, examining it from different angles. What was buried under the next inches of hard clay from which the sandstone had been removed? The grey clay guarded the secret. What was under the piles of loose dirt? I longed to find out, but that, I knew, was not for inexperienced hands to dig into. I tried to see a year ahead. Why couldn't I be here when Brown came back . . . or why couldn't I come back with him?

Back at Paton Ranch, I thanked my new friends for their hospitality. They insisted I stay overnight. I insisted I was full of pressing plans for dinosaur-prospecting and promised I would stop back before I left the country. I couldn't wait to dig into some colorful beds I had seen just over the hill from Brown's buried cache of old bones.

I wound my way back to Shell, stocked up on groceries, and was on my way. Dinosaur hunting, that's the ticket! I couldn't push my rolling home and Harley fast enough, back up that god-awful road toward Howe's. This was no Painted Desert, where bones ranged from scarce to nonexistent; this was a land where dinosaur bones were used for gate stops.

Within a few miles of the ranch, the road forked. The right turn led to the ranch itself, but the left seemed to lead toward the area I wanted to prospect. I followed it for some distance until I came within sight of a sizable wash coming in from the north. Here erosion had left a ridge of brownish-yellow sandstone pitched sharply upward. Below it on the other side, I had a hunch, might be the beds

I sought. The machine was heating up from the hard haul, so I pulled off the trail into a scattered growth of sage. It was but a short walk to the top of the hill.

I was unprepared for the sight below; it slightly resembled the wildest of the Painted Desert, a hell's half-acre-plus of badlands in a hidden basin, color and rock and tumult, all in a mile-long box between me and Howe Ranch. A jumbled maze of hillocks and mounds like painted haystacks; soft shades of vermillion; pinks grading into lavender and reddish purples; interspersed between brighter hues, bands of off-white greys. Dividing the stacks were little gullies, big gullies, little washes, big washes, and away off at the south end a major wash that drained the whole field. The Big Horns, aloof and majestic, rolled up to the east of me with no signs of life. Bright and sere and silent.

I found a place to clamber down, a little gully filled with loose rock, dropping down to meet the rising slope of one of the painted haystacks. The eroded surface was starkly naked, open to inspection everywhere and anywhere. I crossed the rounded slope and dropped to a lower one. This, in turn, led down to a mere mound of lavender and grey clays.

The hard material underfoot crunched its way into my attention. A coarse gravel, the chips a deep greyish purple, darker than the clay. My eyes on the ground, I came upon some large stone spools, rough and weathered, chipped and splintered, of the same hard material that crunched underfoot. They looked like scattered vertebrae of some great animal. Could I have stumbled on the weathered core of a dinosaur's backbone in my first ten minutes in this basin? It was too easy. There were wheelbarrow loads of the stuff in sight. But, judging from the monstrous vertebrae, this must be only a small part of the entire skeleton. What was here seemed to be, on closer inspection, parts of the back and hips. There were no ponderous limb fragments, no flat pieces of ribs, no small vertebrae that might have come from neck or tail, no evidence of skull. Small bits carried down a gully into a wash a hundred feet away told where much of the skeleton might have gone.

As I stared down the wash, a bit of sunlight twinkled back at me, like a diamond. Likely a bit of a broken bottle. I moved, and the sunlight winked out, but I made out the outline of a polished, egg-like shape from whence the twinkle had come and went after it. An irregular stone, half the size of my fist, colored quartz or jasper. I twisted it in the sun, and the colors flashed back at me, highlighted, like polished granite. What could it be? No stone was ever polished like this by wind or water. It had had two opportunities of acquiring polish; it had apparently spent much of its time lying here in the mud and clay near a dinosaur skeleton . . . A near-forgotten newspaper clipping came rollicking back from my memory bank, a clipping about a tray of rocks in a drawer in the American Museum of Natural History, a reporter seeking an explanation of a box of polished rocks, and Dr. Barnum Brown explaining, "These are dinosaur gizzard stones. Some of the dinosaurs picked them up, as chickens pick up gravel, and swallowed them to help grind their food."

"The gizzard stone of a dinosaur."

The quotation marks are correct; I said it out loud. I may have shouted it. A millions-of-years-long time-wave swept over me there in the desert, and I could see the great reptile picking up this stone, attracted by its bright color, carrying it perhaps his life long, dying here. A beast as large as this—the vertebrae were larger than stove pipes—might have died filled with bucketsful of such stones! I tramped back and forth, back and forth, across the purple clays below the bones, picking up a second, a third, a fourth shiny pebble, all bright and colorful. I couldn't get enough of looking at them, handling them, accustoming myself to their shiny brightness and to my dumb luck. I pulled myself back from the past, projected myself into a convenient future. I could see myself now, hear myself toying with them in my pocket, taking out a stone and handing it to a puzzled bystander who wouldn't know how he got in this dream-script and who would ask what in the world it was.

"This," I must say nonchalantly, "is a dinosaur gizzard stone."

Greed interrupted my daydreams, and I resumed the hunt on down the wash. By sundown I had all the polished stones I could pack in my pockets. With no bright sunlight to highlight them, they became harder to find. I sorted out the best, abandoned the stones that had weathered so long as to lose their high polish, and selected the best vertebra I could find. By this time night had set in, and I set out in search of the "Harley-and-houselet" parked in the sagebrush at the head of the wash.

Restoration of *Protosuchus* as a living animal. Photo by H. S. Rice and Thane Bierwert. Courtesy Department of Library Services, American Museum of Natural History.

Barnum Brown. Photo by Barclay-Long.

6

"DR. BROWN will be glad to see you," Mrs. Lord said. The secretary to the Department of Vertebrate Paleontology smiled warmly at Father and me from behind her desk. "You may go right on in; to you, he won't be busy."

Father had come into the city to attend an entomological meeting. I was here, I hoped, to meet Dr. Barnum Brown at last. Some months had passed since the day I had found the gizzard stones.

Father and I went into a long corridor, filled with strange odors and sounds of muffled thumping from a room across the hall. On the glass of the first door we came to was the name and title, "Barnum Brown, Curator of Fossil Reptiles."

It was a very cold doorknob I laid a hand to, and I was suddenly troubled with nervous knees. I had "boned up" since that way-back day I had stumbled onto the strange skull at the edge of the Painted Desert. I had read about Marsh and Cope and their work in the great days of dinosaur discovery and bone-hunting in the last century, and I had poked a bit into Barnum Brown's achievements in the field. Marsh and Cope had long since gone on to happier hunting grounds, gone on at the time a young Barnum Brown had just started out in the field. Beyond this cold doorknob was the third and last and biggest of the three Kings of Dinosauria. And here comes a country boy who had stumbled blindly over tiny bits of the same ground and who brought in his pocket a few choice gizzard stones Barnum Brown hadn't had time to find . . .

Father knocked while I was frozen on the doorknob, and a friendly voice called out, "Come on in." We entered a long room, a lengthy table running down the center, glass-doored cabinets lining one side of the room, a roll top desk beyond the table. I was vaguely conscious of a dinosaur picture on one wall and trays of bones on a table, but my eyes were held most by the man at the desk.

Father spoke and shook hands, and they chatted idly a moment. Brown looked exactly as I had seen him in photographs. The pince-nez, the shining bald head with its fringe of grey . . . these were familiar. And yet, in a strange way, he looked different. Was it the finely tailored business suit, the look of his well-manicured hands, the perfection of his whole ensemble? Suddenly Father was introducing us.

I reached for Brown's extended hand; the great and long-awaited moment had come. He mentioned something about Winslow, Arizona, and the wrinkles around his eyes were smiling. It came to me in a moment of panic that I, too, should speak. But beyond the formalities of introduction, I was without words. Fortunately Brown kept on chatting. Familiar names fell on my ears . . . "Hubert Richardson . . . the Triassic of Arizona . . . Cameron."

He turned to the bench above the cabinet drawers and picked up a snouted skull from a tray there. "You just might find this interesting," he remarked. "My" skull, my very own skull. It was, of course, a plaster copy, a cast made by joining the two stone halves together and filling the cavity with plaster. The cast had been painted light grey, the color of old bones. Brown's sensitive fingers seemed almost to caress the wrinkled surface. I could see now why he had been so concerned about the second half; like Father with his moths, Brown's fossils were his children, his babies.

"We were afraid at first there were no direct contacts between the two pieces," he said, "but Mr. Falkenbach, of our lab force, found a union along one edge. That was enough."

He returned the specimen to the tray and turned to me, his eyes twinkling. "I wouldn't be surprised if you'd like to see the lab."

Would I like to visit the laboratory? I knew, of

course, there must be a place where fossils were prepared for exhibition, but getting a tour of the place was more than I had counted on. Dr. Brown led us out into the corridor and through the door of the opposite room, the one from which the muffled thumping sounds had come. As we entered, I was aware this room was the source of the odd odors in the hall; mixed with more exotic odors was a strong smell of shellac. The muffled thumping turned to loud clanging. We were in a small room which led into two somewhat larger rooms. An overhead track carrying a sizable chain hoist ran through all three rooms. In the first room, over by the big windows, a blacksmith's forge stood at the end of a long bench loaded with metal-working tools. A short, stocky man stepped back from an emery wheel on which he had been grinding a metal bar, turned the partially mounted skeleton of a fossil animal I couldn't identify, and fitted the piece of metal along a bone.

"I want you to meet Charlie Lang," Brown said. "Charlie is our artist in iron; he's responsible for some of our finest mounts." We shook hands and shouted a few pleasantries at each other; a worker was pounding a heated bar, another was sharpening a chisel, and the din was considerable. Lang discussed bits of his morning's work with Brown and then stepped over to a gas furnace to pick up a bar heated to a cherry red. "Pardon me," he said with a smile, "but I've got to beat this out while it's hot."

We moved on in the direction of another noise source in the next room. A big, white-haired man in a spattered smock seemed largely responsible for the racket, ably assisted by an assistant quite as large and as sturdily built. The table at which they worked was cluttered with containers filled with smelly gums and mixtures and large trays loaded with chunks of rock and bone. The two men were hammering away with chisels on a huge rock which stood upright on a bench, supported by two-by-fours.

"Otto, I'd like you to meet R. T. Bird; I believe you already know his father," Brown interrupted. He had to raise his voice to be heard. Turning to me, Brown added, "Mr. Falkenbach here is the man who made the skull cast of your stegocephalian."

My fingers were gripped in a brawny hand. The curator went on with an explanation of the work in progress. "This armored dinosaur of a new type is being cut out of solid rock. The stone is very hard, and Otto and Jerry here are practically sculpting the bones out of solid rock. Pete Kaisen and I brought this back from Cashen's Ranch down on the Crow Indian Reservation south of Billings."

Brown turned to the table with its trays loaded with lumps of rock weighing five to twenty-five pounds apiece.

"Some of it was weathered and broken up like this," he continued. "Otto has to cement them together as he goes along with the work. When it's completed as a double-panel mount, it will show on one side the plates the creature carried on its skin, and on the other, the internal skeleton."

Mr. Falkenbach showed Father and me a small model of the dinosaur as he appeared in life. It looked for all the world like a giant horned toad, except for its slightly more erect posture.

In the next room men at small lab tables were surrounded by more of the same paraphernalia—trays, gums, tools. They were working on blocks that contained parts of another animal. But in contrast to what we had just seen, here was a specimen in which assembly was more a matter of cleaning and repairing rather than sculpturing. The encasing matrix was soft and shaly, the fossil bones greatly fractured. I met Carl Sorensen here, one of Brown's top lab workers. He handed Brown two bits of bone. "These seem to fit together," he said, "but I'm not sure." Brown manipulated the pieces deftly, holding them close under a shaded lamp. The contact, it seemed, bore an important relationship to the rest of the specimen. He bent to the bone fragments, trying them this way and that way. Finally, he found a fit and confirmed another made by Sorensen.

Mrs. Lord invaded the lab with a message for Brown. "You're wanted on the phone." As Brown left she told us, "He'll be back in a minute. Don't fail to have him take you through the dinosaur hall."

It was years since I had seen the dinosaur hall. In boyhood I had made many trips to the museum with Father, but our tours were generally confined to the insect hall and habitat groups of modern animals.

Brown came back in a moment. "If you'd care to," he said, "we'll have a look at some of the mounted animals."

Father and I followed him out into the hall and to the elevators. The trip through the laboratory had shed a new light on Dr. Brown, the suave and polished curator; the rock-worker and digger of old bones was represented in the persons of his aides, in the hammer blows of the rugged Otto Falkenbach, in Charlie Lang's roaring furnace, in Carl Sorensen's fiddling and fitting with bits of old bones.

On the fourth floor we entered a tremendous room. Across its wide tile floor and against its high white walls marched a tremendous menagerie of mounted skeletons that seemed alive and in movement. One reared in defiance of an enemy. Another seemed about to lunge at an adversary, his heavy frame and sharp horns set to deliver the impending blow. Other bone giants, indifferent and oblivious, lumbered or waddled about in situ on dinosaur business, complacently unconcerned, too big or too well-protected to fear attack. By way of contrast, a glass case in the immediate foreground contained a small creature, not much over a foot in length, embedded in a block of rock. The matrix was a familiar bright red.

"*Protosuchus*," Brown said. "The little ancestral crocodile. He came from the Upper Triassic, east of Cameron, not far north of the dinosaur tracks."

Again I could see the grotesque formations of that fantastic part of the world. Surely none of the passing visitors, looking at this same specimen, could feel as I felt. I pitied those who had never seen Dinosaur Canyon.

We paused before a large wood and plaster base. Two huge clawed feet were poised here, one in advance of the other, as if walking. The leg bones looked like those of a giant rooster. The ankles were two feet above the ground; the knees well above our heads; we had to crane our necks to look up at the great hips to which they gave support. The rest of the creature reached to the high ceiling, its skull with a four-foot jaw filled with dagger-teeth leering horribly at the nearby lighting fixtures. A long balancing tail dropped away in the opposite direction, sweeping low toward the floor. Forty-eight feet from head to tail. "*Tyrannosaurus rex*," the sign said, the greatest, the mightiest meat-eater ever to have walked the earth.

"This fellow came from the Hell Creek beds in Montana," Brown said, and went on to tell about its discovery. "I searched the badlands for a month without finding a thing. I had given up; then I decided to give it one more day. Turned out to be my day; he's still the biggest *Tyrannosaurus* ever found."

The tale went on. "After that, it took two seasons just to get the skeleton out of the badlands.

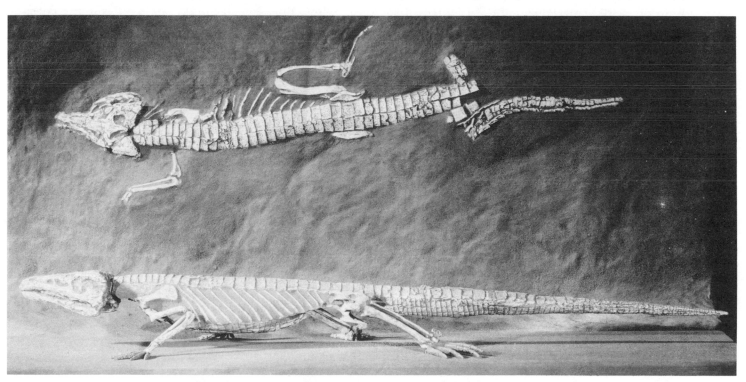

Skeleton of *Protosuchus*, the ancestral crocodile. Above: dorsal view of the original specimen. Below: side view of a reconstruction of the skeleton. Photo by Irving Dutcher and H. S. Rice. Courtesy Department of Library Services, American Museum of Natural History.

Back in those days we had to do all our moving by horse and wagon, and the place where the bones were found was miles from the nearest railroad. Even after we got it back to the museum, our troubles were far from over. Months were required to remove the bones from the matrix. The skull and limbs worked out easily, but the pelvis and some of the connecting vertebrae were encased in a hard iron concretion, which Pete Kaisen had to take out in the yard and knock off with hammer and chisel."

We moved down the hall, past two tall creatures with long, slender heads terminating in duck-like bills. One stood on his hind feet; the other rather as if to drop on all fours to investigate a bit of simulated plant life on the plaster base. Both had long tails, hoof-like nails rather than claws, and a type of grinding teeth clearly meant for chewing leaves and grinding brush. The sign labelled them "Trachodonts."

"These lived in great numbers in Cretaceous times," Brown said. "In Alberta I've found them by the hundreds in a single season . . . though the first one was found right here in New Jersey."

We moved on down the hall, from creature to strange creature. I would come back and digest this more slowly, but for now I could only taste the hour. I recall a nice little fellow labelled *Camptosaurus*. "If you had him around for a pet," Brown mentioned, "you might have called him 'Humpy' . . . out of respect for his scientific name." I didn't

mention how surprising "Humpy" was; like many another viewer in the great hall, I still had the idea that what made dinosaurs interesting was their great size. I hadn't learned yet that only the big ones were big. Dinosaurs ranged all the way down to the size of turkeys and smaller.

We crossed the lower end of the great hall and came at last to *Brontosaurus*. Before me rose a great hind leg, towering toward the ceiling and surmounted by a pelvis large enough for a tree-house. The other hind leg was more than six feet away. "The skeletons we partially uncovered at Howe Ranch were of this type," Brown told us. "Sauropods . . . dinosaurs that spent all their time on all fours."

We took a stroll around the creature's front end and by and by came upon the second set of great legs and within better view of the great neck and tiny head. Hardly larger than a bushel basket, it drooped low to within eight feet of the floor. The entire length of the animal from the head to where the tail just touched the base was some seventy feet.

"The Howe Ranch skeletons proved too large and too many to take up this year," Brown added, "with the limited means we had. But we expect to go back next year and try again."

This was the magic moment when a young fellow I knew well should put in his plug for a place on the project. I had discussed the idea with Father, and he concurred heartily.

"Is this the fellow who picked up gizzard

Skeleton of *Apatosaurus* (*Brontosaurus*) collected by Barnum Brown at the Bone Cabin Quarry. The skull on this mount is incorrect; the skull of *Apatosaurus* was actually more like that of *Diplodocus*. Courtesy Department of Library Services, American Museum of Natural History.

44

stones?" Father inquired. There went my moment.

"There's little doubt this fellow is one of the dinosaurs that picked up rocks to help grind his food," Brown replied. He called attention to the teeth in the creature's skull, insignificant and worn stubs set well front in the jaws, the absence of grinders indicative of the probable need of gizzard stones.

"By the way, I just happen to have a few odd rocks I picked up in Wyoming, where you were tracking dinosaurs. Would you care to have a look at them?"

Skeleton of *Tyrannosaurus*. Courtesy Department of Library Services, American Museum of Natural History.

Two skeletons of the hadrosaur *Anatosaurus*. Courtesy Department of Library Services, American Museum of Natural History.

7

THE LETTER READ, in part, "Carl Sorensen is leaving for Wyoming, in charge of an advance party, on the twenty-fifth of May. If you care to join the expedition at Howe Quarry . . ."

The year was 1934. The letterhead was that of the American Museum's Department of Paleontology, the signature below the typing, "Barnum Brown." The message I had hoped for for five long months was in hand. If I left Florida in the next five days, I should arrive at the Howe Quarry at about the same time as the advance party.

Impatience rode with me on the way west. About noon of the day Sorensen and his boys were to put in appearance, the Harley and I struggled up the rise of the rough little road leading to the Howe Ranch. Four patches of white dotted the little valley beyond the corral: tents, stark against the green alfalfa, like handkerchiefs laid out to dry in the sun. The New York party was already in. I bounced and rattled down the rocky grade toward the ranch gate, scanning the hill above the corral for signs of men or movement, dismayed that I might have arrived too late for the opening of Brown's cache of bones.

A wisp of smoke rising straight from the ranch house chimney indicated perhaps everyone was at dinner. I opened the yard gate, drove through, paused to close the gate with great impatience. A green Buick touring car was parked under the cottonwoods near the corral. Next to it, a Ford truck with closed body. I drew up beside the cars, a few feet from the nearest tent. Hearing voices, I raised the tent flap and looked in. Carl Sorensen, seated at the head of a new pine table, saw me at once.

"Glad you found us," he said. "Brown said you'd probably drop in about the time we got here. Let me introduce the rest. Ted Lewis . . . Dan Thrapp . . . Bill Frutchey."

Lewis, tallest of the three and the only one who might be called well-groomed, stood up to shake hands, picking up a new pearl-grey Stetson from a nearby chair as he did so. "The noticeable lack of industry which you may detect," he said, "is, we hope, temporary and due to waiting to see Frutchey wash the dishes."

The boy Frutchey, a stocky, black-haired youngster still seated at the table, laughed loudly. "They've all got it in for me," he said. "I'm the cook, and nobody appreciates me."

Thrapp didn't enter into the conversation, looking on and listening with quiet amusement, an unobtrusive, easy-going sort.

Frutchey put an armload of dirty dishes on a box next to a gasoline stove outside. I noticed he wore knee breeches and what appeared to be golf socks with a pair of heavy, square-toed shoes. Lewis came out of the tent behind him, smiling satirically. "Bet Frutchey's the first camp cook you ever saw wearing ski boots," he remarked.

For my camp, I picked a spot just north of the tents, a stone's throw from the hill where the dinosaur bones lay buried. Each day I'd have only a few steps to walk to reach the winding path up the steep flank on the little hill. I hurried with my unpacking, still half-afraid I might miss the uncovering of the skeletons. Lewis and Thrapp were nowhere in sight, and I wondered if they might have already gone up the hill. Then I caught sight of Carl down among the cottonwoods, nailing up a small shelf to hold a wash basin. Nearby, Frutchey was washing dishes. Carl finished his job, and started walking leisurely toward the last tent in the row, and I hurried to intercept him. "Well, I'm ready," I said, "anytime you are."

"Ready? For what?"

"Why . . . to start uncovering bones."

Carl laughed. "Don't push us; we just got here," he said. "Now, let's see . . . tomorrow's Saturday. We'd just as well spend the weekend resting up from the trip and doing what needs to be done around camp." I must have looked both downcast and embarrassed. Carl laid a hand on my shoulder and shook me playfully. "Don't take it so hard," he said. "You'll see enough bones before the summer is over."

I inquired about Brown and Pete Kaisen. Brown, Carl said, would be along about June twentieth. Kaisen had been taken ill and would miss this summer's digging entirely. The conversation turned to the rest of the party, and I learned Ted Lewis was finishing his doctorate at Yale and had worked in the field with Brown before. Thrapp was from the laboratory force of the museum. Frutchey, a college student from New Jersey, had applied for apprentice work with the expedition, and Brown had agreed to take him on if he would serve as cook.

Carl's choice of a spot to pitch the tents was ideal. The nearby row of cottonwoods furnished cool shade close to the hill on which the bones were buried. Enough distance lay between us and the Howe ranch house to give us both privacy. Fresh water from a small but steady spring that flowed into wooden troughs just below the house furnished abundant good water. A large garden between the Howe's front yard and the line of cottonwoods would furnish vegetables for all of us for the summer. The wretched road to town was the only drawback; it was badly washed out from the winter rains; each trip to Greybull for supplies would consume many tedious hours and a lot of automobile.

Supper, when it came, was the event of the day. I wondered how I could possibly wait until Monday to start digging bones, but apparently this thought entered no mind but mine. Lewis continued to tease Frutchey about his ski boots, and even Thrapp added his bit. "Frutchey's going to donate those boots to the museum when this trip's over," he said. "They're to be put in a special case beside the bones."

I beat the sun up, but barely. I met Lewis in front of the cook tent, where he stood studying a great rising bench of sandstone, red among the green cedars of the rocky slope. "Chugwater," he said.

"Chugwater Triassic . . . that particular red bed." Lewis had seen my puzzled look and ex-plained, "The formation is named for the locality where it was first described."

So the bright red rock was Triassic, belonging to the same series as those in the Petrified Forest and Dinosaur Canyon. I was learning a little . . . if only I didn't betray the vastness of my ignorance faster than I learned. And this bit of learning was named for a town called Chugwater.

"I'd like to climb up there this morning and see what the rest of this part of the world looks like," Lewis went on. "Like to go along?"

I welcomed the suggestion; I could learn a lot from this man. As soon as breakfast was over, we all set out together . . . all but Carl Sorensen, who had letter-writing to catch up on. An old sheep trail led up over the hill behind the ranch house. We followed it to the head of a small canyon, where it steepened sharply. At first there was good-natured banter about Frutchey's ski boots, but as we strung out and the trail steepened we diverted our energies from wit and banter to work and heavy breathing. The path wound into a low growth of cedar and across a slope of the red sandstone. Lewis and I, with Thrapp right at our backs, reached the top of the sandstone bench. We didn't stop but kept on for higher promontories projecting from the curve of the limestone ahead. Frutchey dropped from sight in one of the turns of the winding path below.

The air grew cold and sharp with the increased altitude. We were up now in a region of cedars and limestone, climbing for the rim of the sky, tiny ants crawling over the sloping belly of the giant who slept beneath the mountain. Below us the wide Big Horn Basin stretched away toward the Rockies, forty or fifty miles off. Sheep Mountain, north of Greybull, blocked them partly from view. Finally Lewis paused on a windswept ledge, an ideal vantage point from which to see the formations as they lay exposed on the eroded mountain slope. The flaming red sandstone was a spectacular zigzag band, rising and falling with the washes and canyons cut through it. Above it stratigraphically, but lower on the slope, were beds showing a ruddy blush. These pink bands were duplicated along the flanks of Sheep Mountain. Together with the other beds, they had been uplifted by the arching of the same underlying limestone that lay beneath our feet. The series of scalloped edges was striking.

"The pink beds are Sundance," Lewis said. "Next to them are the Morrison beds . . . dino country. And next to those are the Cloverly." A question popped into my empty but receptive mind. "How about those bright beds at the Howe Ranch, just above the hill where the bones are?"

"Those are Cloverly too . . . described from Howe Ranch where there used to be a Cloverly Post Office."

Sundance particularly entranced me as a name, poetry appropriated by science. Later on I learned a "Sundance Sea" once covered a great area in this part of the world. A long time later, I drove through the town of Sundance. The flat, baked little town in an over-dried and baking world didn't do a thing for the poetry of the name. That's the way it goes, sometimes.

"What happened to Frutchey?" Thrapp asked. We looked back down the trail; there was a tiny figure seated on the rising front of the Chugwater. Lewis chuckled.

"Frutchey is saving his ski boots for science," he said.

Monday after breakfast Carl announced he was ready to open the quarry. I liked the term, ever since I had first heard it used by Brown; it suggested clanging hammers and breaking rock. We all followed Carl into his tent and helped drag out a heavy pine box. Inside was a carefully packed collection of tools. From a cardboard box on top, Carl picked up a handful of small instruments with wooden handles. "Every man will need a crooked awl," he said.

The tool was much like a shoemaker's awl but with a curved point and flattened at the end. We each also got a straight awl, a bit more rugged than the shoemaker's awl. This was followed by a stiff little whisk broom. A flat, knife-like digger came next; Carl explained this was for undermining

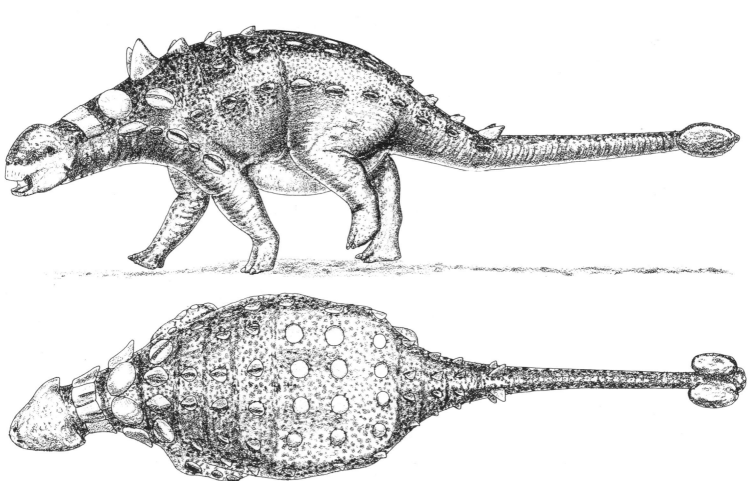

Two views of *Euoplocephalus*, an ankylosaur from the Late Cretaceous Judith River (Oldman) Formation. Restoration by Kenneth Carpenter. Courtesy Kenneth Carpenter.

bones. Two small paint brushes, one an inch wide and another half that, all but completed our set of tools. Carl searched through the packed carpenter's tools, sledge hammers, cold chisels, past a heavy chain hoist, finally came up with a long package in brown paper. It was full of wooden handles about twenty inches long, hafted like small pick handles. He passed these around, followed by light steel pick heads about ten inches long, pointed on one end and flattened to a one-inch blade on the other. We slipped these on the handles and hefted them; they were a delight to handle.

"Don't let anything happen to that prospecting pick," Carl warned. "They can't be bought in the local hardware store. They're made up on special order, to Brown's specs."

Ordinary picks and shovels from the truck completed our equipment. Carl poured a quart of shellac into a gallon can and thinned it down to a watery consistency with alcohol. We set out for the quarry, ready to dig bones.

At the crest of the hill we gathered about the long mound of earth at the east end, undisturbed since I had been here months ago with the Patons. Carl and Thrapp began to scoop away at the outer edge of the mound in a matter-of-fact manner, as if they were merely shoveling dirt. Lewis looked about for a place to set his camera so he could join the ceremony. I stood back, concentrating on keeping out of the way. Presently Carl's shovel struck a soft, yielding substance. This, of course, could occur at any disinterment . . . but it was only old newspapers, put there to protect the bones months ago.

"Take it easy, now," he said to Thrapp. "We're almost there."

They scooped along the paper layer, following the edge of the mound. Lewis squatted down and began to brush away the remaining dirt with his whisk broom. I itched to be asked to help.

"What can I do?"

Carl motioned me to work with Lewis. Fifteen or twenty feet of the mound was cleaned off, along the edge of the hill. Carl and Thrapp laid aside their shovels, and we all began to pull away at the paper matting. The scant moisture that had seeped through the soil during the winter months had dampened the paper on top, but the sheets underneath were dry and firm. I lifted one of the bottom sheets and saw a bit of the bone. Black, like what I had seen along the face of the cliff the year before, it was badly broken and shattered, but the fragments lay undisturbed. Another lifted paper re-

vealed fragments of more bones. Vertebrae, parts of a dinosaur's tail, half exposed and half embedded in the soft clay. Beside me, Ted Lewis dusted off the exposed surfaces with the larger of his two paint brushes.

"Articulated caudals. Sauropod," he said. Fine way for a Yalie to talk, in front of a simple country boy! "Tail-stock from the big fellows who walked on all fours." I felt better. Lewis went on to point out that the neural spines were complete, and that the chevrons, the pendant, spine-like bones that once supported muscles beneath the vertebrae, were also in good shape. My vocabulary was shaping up a bit. Nearby, Carl and Thrapp uncovered a large flat bone, which they immediately pronounced a scapula. I knew that one: the shoulder-bone of a sauropod dinosaur. If, now, I could only bring my digging and excavating along with my expanding vocabulary . . .

We spread out along the face of the outcrop, each individual continuing to lift and discard papers from the layer of bone. When the last of the protecting mat had been removed, we saw we had two complete hookups of caudal vertebrae several feet in length, each leading into what appeared to be the hips of an animal. With them were several ribs, the single scapula, and a few other incomplete bones. The last, Brown had found partly weathered on the hill. The entire lot were as incoherently mingled as if they had been two old cow skeletons found side by side in an open field, but it was obvious still more bones ran back under and into the layer of clay. It seemed we had only well begun to work when Carl glanced at his watch and announced it was time for lunch. Frutchey had, quite properly, slipped away some time before. "This afternoon we can start stripping off overburden," Carl announced.

"Overburden" was just the layer of clay lying above the part of the bones not yet exposed. It was only eighteen inches thick until it joined the remaining sandstone caprock not removed by Brown the year before. To keep the already uncovered bones free of dirt that might fall on them as we shoveled, we spread fresh newspapers on them. When we had cleared a few yards and were down to the level where we might encounter bone, Carl had us lay down our shovels. "We'll take up the papers now," he said, "and start following the surfaces of the bones with our awls and brushes." He took his crooked awl and started working on a bone that was partly exposed at one end but continued in under two or three inches of remaining clay. "You can feel the bone with your awl before you see it," he said.

Lewis and Thrapp were already on their knees . . . prying, not praying . . . farther along the outcrop, but I watched Carl a bit before attempting exploratory work on my own. He proceeded with great care; the bone itself was greatly shattered. When he had pulled back enough matrix to clear several inches of surface, he gently brushed away the last of the debris from the bone with his smaller brush.

There was the sound of footsteps behind us, and a shadow fell across the quarry. Rancher Howe, wearing the same battered slouch hat he wore when first I met him, came up and seated himself on a pile of clay.

"I see you're diggin' away at my old bones again," he said, taking note of the specimen Carl had just dusted off. "But they're so busted up; how do you ever expect to get them out, now they're in so many pieces?"

"Don't worry; watch this," Carl said, reaching for the shellac can. "This might be good to know sometime, if you're ever thrown off a horse and get broken up. We'll fix these so you'll never know they were broken."

He dribbled the shellac from his flooded brush into the cracks and fractures of the badly broken surface. The dry bone bits drank in the yellow fluid thirstily. "When that sets," Carl went on, "the bone won't be in any danger of falling apart before we're ready to take it from the ground."

I tried my hand at working on a partly exposed rib; following a rib's simple outline should be easy. It was fascinating to watch the bits of clay fall from the awl point, to see more and more of the hard, black surface appear to view. I seemed hardly started when Carl said it was supper-time. I would rather have gone on until dark, but I knew such a suggestion wouldn't have met with approval of the others. Again Carl had us cover the exposed bones with newspapers for protection against the possibility of rain.

That evening Lewis sat down at the table after supper with a book on paleontology, full of drawings of sauropod skeletons. I sat down beside him, asked a question or two and ended with an hour or more of instruction. Mostly, I learned how much I didn't know about reptilian anatomy.

Next morning we went back to removing another section of overburden. One of the boys,

wielding his prospecting pick, struck a bone elevated somewhat above the rest, but little harm was done; the matrix was harder than where we had begun, and the bones buried in it less weathered as well. As a result of this near-mishap, however, we returned to using our curved awls. The bones, for the most part, were a genuine bone-hash; when we ran out of one, we knew that an inch or so away we would find another. The morning's big find was a huge femur, the end of it the size of the Howe Ranch gatepost.

Dinner-time again came on with amazing speed, and dragged by. For all our work, the appearance of the quarry had been changed little. At supper-time I got off my knees disconcerted to see how little we had accomplished during the afternoon. "One day's work doesn't leave much of a mark," Carl said.

In the evening the weather, which had stayed cold, turned threatening. Sometime during the night it began to rain. I woke up to look out on a grey dawn with the chilling sight of snow on the mountain above; the white blanket, in fact, extended down to within a thousand feet of the camp. Our quarry on the little hill looked wet and forlorn indeed; there would be no bone-digging today.

About halfway through the morning, we ran out of coal and took to the wood pile. Wood is, after all, the wonder fuel; it warms twice . . . while being chopped and as it burns. While we were engaged in the first half of this energy-providing process, Milo Howe, the rancher's son, came over from the corral to see how we were making out. A powerfully built, muscular man in his early thirties, he looked beaten and dejected.

"I'm worried sick about my sheep," he said. "We sheared them only last week and turned them out on the mountain; we never dreamed we'd get this kind of weather so late." He wandered away but came back in a bit to stand shivering beside our hot cook stove. "This wet snow may have killed half my sheep," he groaned. "There are dozens of them dead and dying in the lower draws above the house. I'm on my way now to the upper camp." A couple of herders, he explained, were stationed at this camp near the upper pasture.

Carl, Thrapp and I climbed up with him through the scrub forest to see if we could help. We soon reached the snow line. The white blanket, wet and cold, lay everywhere. We came upon a brush-lined draw and saw ten mounds of snow among the rocks . . . but it wasn't rocks that gave shape to the rounded mounds; steam rose from some of them! From others, there was no sign of vapor. I pushed the snow from one mound with my foot. The dead creature was naked, the scars of its recent shearing still fresh and red against the cold white skin. The sheep that were still breathing were more pitiful. In the next draw we came upon more dead and dying sheep.

At the top of the mist-shrouded mountain, at the upper camp, we found more of the same, and nothing we could do. Sheep still alive couldn't be moved; there was nowhere to move them. Naked, coming down with pneumonia, snuffling and gasp-ing for breath, the barely living lay among the better-off dead, a sorry sight indeed. There was nothing we could do to help; we were all three of us glad to get back to camp, to our buried dinosaurs. Tragedy had befallen them, too. But that was a long time ago.

It was two long days before we got back to rancher Howe's old bones, until the slippery, squishy clay had dried out enough to work in. Milo Howe's losses in sheep ran into the hundreds, but some of the younger and stronger ones recovered. In the quarry, our forethought in covering the bones with a generous layer of newspapers paid off.

8

CARL PUT DOWN the shellac can and turned to Lewis.

"If you ask me, there's something of a bonus in this hillside, over and above the two dinosaurs Brown sent us out here for."

Inexperienced and innocent as I was in this work, the same thought had been creeping up on me. More dinosaurs.

"We've already turned up five scapulas, and here's another string of caudals," he went on. "I'll be badly mistaken if there aren't four or five more or less complete skeletons. I had that notion Sunday, looking at what we've turned up and at Brown's pile of fragments." The bones at the end of the quarry were the ones Brown had picked up from the detritus at the foot of the hill, weathered from the bone outcrop. Checking the evidence, it did indeed seem we could make a good case for four bodies.

Excavating was resumed with new vigor. What lay under the next twenty-five feet of the still undeveloped part of the quarry? I took to working regularly with Carl, digging on, hoping to find in the balance of the bone layer the rest of the elements to complete the partial skeletons already exposed. Instead, we came upon the vertebral column of another individual that led to another pelvis. Beside us, Lewis and Thrapp ran into an assorted mass of packed bones that seemed to have no bottom. At the same time Frutchey, working sporadically at the north end of the quarry when not busy in camp, came upon vertebrae of a great neck, eight to ten feet long.

I could hardly bear to take time out for eating and sleeping; not in wildest dreams had I pictured such a mass of fossils, such a paleontological jackpot. But Carl was worried.

"I don't know what's to come of this," he groaned. "We came out here to collect two dino skeletons; now we have more on our hands than we can count. Couldn't get 'em all back if we could drive them home."

Carl's letter to Brown, telling him of the apparent great expansion of the project and assuring him of an eager welcome on June twentieth, was promptly answered, in what Carl claimed was real Brownian style; we got three new men. Dr. Laurence F. Rainsford and his son, Laurence, from my home town of Rye, and a Dr. Green, from Princeton, joined us. Two more tents went up, and our camp began to take on the aspects of a canvas village. Carl put the new arrivals to work where they would do the most good, and work went merrily on. Our new arrivals made a fine addition to our party, socially and on the labor front. And with Lewis and Frutchey verbally rampant as usual, something of a carnival spirit prevailed most of the time. But the additional manpower failed to jar Carl out of the doldrums. "We've got to start taking up these bones," he said, "They're getting underfoot. But I can't do a thing 'til Brown gives the word."

I was growing curious about the next step in the operation. It's one thing to disinter a tangled mass of dinosaur skeletons; it's quite another to untangle them and ready them for shipment. The problem was keeping a record of their relative positions. Though to a degree disarticulated and disarranged, many bones held a certain association indicating their place in one skeleton or another. Others were tumbled like huge piles of giant jackstraws. I felt the position of each bone as it lay in the quarry should be noted somehow and broached the subject to Carl. "We'll have to have some sort of a quarry chart," he said.

On July third, Lewis drove in to Greybull to

53

pick up Brown . . . here at last. It was after dark when they arrived in camp, and we all crowded into the cook tent to greet the long-awaited head of the expedition. He didn't have much to say but sat puffing a curved briar, listening to Carl's report. The gasoline lantern cast a bright glow on the faces of the audience, a glow that flared from time to time when Brown nodded and smiled. I was in a far greater state of excitement to witness his reactions to sight of the bones than he was to see them . . . but then, after all, this was my first dinosaur graveyard.

Of course we started for the quarry next morning without waiting for our breakfasts to settle. We had all worked like dogs digging bones here, and we wouldn't have missed Brown's reaction to the first sight of what we had done. We got it.

"A dinosaur treasure trove," he said. "An absolute, knockout dinosaur treasure trove!"

Then he got down among the bones, examining a piece here, another there. His hands caressed these bones as lovingly and gently as they had my stegocephalian skull that first day I had met him in the museum office. The fact that these bones were rapidly getting in the way of bones we had yet to dig up didn't seem to perturb him.

"We'll go right on removing overburden for awhile," he told Carl, "but something's got to be done about recording relationships."

I spoke up. "How would it be if I stretched strings across the quarry . . . say a yard apart . . . and worked up a quarry chart? Fast as I draw in the position of each bone, we could take the strings down, and then get the bones out of the way."

"Fine, R.T.," Brown replied. "Get right on it. If you need help with the strings, Thrapp or any one of the others can lend you a hand." He reflected a moment. "Then, do you think you can later draw in additional bones that crop up as the surface bones are removed?"

"Sure; no sweat."

Brown studied me reflectively. "Come back and say that over, after the job is done," he said.

I got on the job at once, hoping I hadn't promised more than I could deliver. It was evident Brown wanted to see all the bones we could uncover this season.

"You can count on Brown never to get enough," Carl said that evening, "when it comes to dinosaur bones." I understood perfectly; I was just as greedy.

Brown's arrival loosed a lot of fresh activity

about Howe Quarry; Milo and his heavy team of grey horses were added to our group, set to removing more heavy caprock from the top of the hill, pushing the territory farther ahead of the advancing picks and awls and brushes. Don Thrapp and Laurence Rainsford worked with Milo at this dull and rugged task. Brown turned out, tools in hand, taking a place among the squatting bone-diggers. I missed my old job and the fascination that went with it, but I couldn't possibly chart bones and dig them at the same time.

Howe Quarry suddenly jumped into the national limelight. It had attracted little notice at first, being so isolated and so far from any sizable city, and only a few local ranchers had drifted in to look and marvel. The Greybull papers picked up the story, with Brown's arrival, and then Associated Press and the radio people got into the act. Lewis had sent some pictures back to the museum, and these now appeared in a Sunday edition of *The New York Times*; we were definitely in the public eye.

The fanfare was not an unmixed blessing. It was excellent publicity for the American Museum and for Sinclair Oil and Refining, which had generously financed the expedition, but we in the field felt the brunt of publicity's adverse effects. We were unprepared to handle the never-ending procession of cars climbing the ridge south of the Howe ranch house. Nor was the road in any condition to bear up under such traffic.

"This quarry is a wonderful sight!" one visitor said. "I appreciate it. But boy, what a car-killing road!"

"I'm sorry to bother you at this late hour," another apologized, "but it took me all afternoon to get up that doggone road." He had arrived after dark; tired as we all were, we felt obliged to give him and his party a flashlight tour. Then there were those who got lost, taking the wrong turn and ending up at a neighboring ranch. Others turned back when they saw the rocks, ruts, and grades ahead of them and left the region with a feeling of resentment toward us.

"I hate to take Milo and his team out of the quarry," Brown said one evening, "but I guess we've got to do something about that road."

Milo worked his horses on it two days with a scraper, but it was still a rough trail, not a highway. In spite of this, the Howe corral, once oozy with old sheep manure, remained a humming parking lot. Autos of all makes and vintages found their way in to us, from Billings and Butte, from No

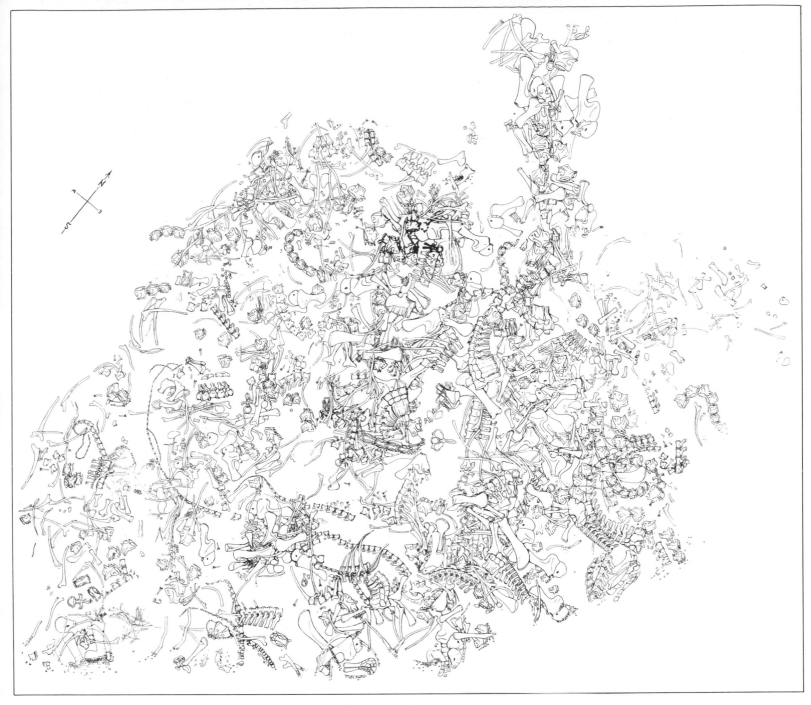

Bird's map of the Howe Quarry site.

Wood and Medicine Bow. Wyoming and Montana licenses predominated, of course, but there was a scattering of cars from everywhere. Some failed to make the final mile; they parked at the foot of the last bad climb, and their owners walked in while letting their radiators cool.

"My wife begged me to turn back," one visitor said, "but now we're here, she's glad she couldn't talk me out of it." Saturdays and Sundays were, of course, the days of record crowds. Some lingered to watch the work go on for an hour or two; some brought lunches and spent the day. The drawback of a continuous but changing audience was that everybody asked questions . . . the same questions.

"How did you know where to start digging?"

"Have the bones been buried long?"

"How did the skeletons get here?"

"What did these animals look like alive?"

"Were these dinosaurs here before or after Noah's ark?"

Brown took great pains in answering all queries, though it is difficult to give care at the same time to work as disparate as cutting a new-found bone from its matrix and treating solemnly a question already answered dozens of times. He would, each time, look up from his work and listen attentively, then return to his work while answering in patient detail. The rest of us, less practiced, couldn't handle both jobs at once as well.

"This site was probably an old water hole," began the answer to the most-asked question, "left over from a much larger body of water that dried up gradually. The bones you see here were those of sauropods . . . the big fellows who walked on all fours . . . creatures whose very size made them largely dependent on an aquatic environment. In all probability they were too heavy to support their enormous weight on their legs for much of the time; they floated or wallowed around along the shoreline, feeding on plant life along the banks. The drying up of the waters drove many dinosaurs to this one spot, where they died, standing on the trampled bodies of dinosaurs who had died here before them.

The "looped" whiplash-like sauropod tailtip found at the Howe Quarry. This tail may belong to *Diplodocus* or *Apatosaurus*. Photo by R. T. Bird.

"The water hole may have refilled from time to time, and the process repeated. Rushing waters, bringing in silt and sand, buried everything. Eventually the bones fossilized, and they have remained in this condition at least the last eighty or ninety million years."

Most of the visiting public were well-mannered, careful, and appreciative, but there were exceptions. The Bone-Kicker Sightseer was the species highest on our list of pet hates. He usually arrived at the edge of the quarry either very bored or very jocular, looking down on the nearest bone. "Is that a bone?" he customarily inquired, either testing or pointing with an experimental kick. In the case of a good, solid limb, it didn't really matter. But even a gentle foot-pat on the head of a small rib would bring it down in a shattered heap. "Oops! Sorry."

The matter was brought to a head by the Vacant-Minded Stumblebum, a species thoroughly detested by our scientists. This cabinet specimen slipped right by me, unseen until I heard Lewis shout, "Get out of those bones! What're you trying to do?"

I looked up, to see the man wheeling wildly, right in the middle of three skeletons; how he got where he was without doing damage, I'll never know. He jumped at the sound of Lewis's voice. So did we all; so did he, poor man!

"Look out for that rib!"

"Keep to your right! Take care when you turn!"

He started weaving toward the edge of the quarry. The end of a small rib bounced from his boot. A scapula crumbled in his path. The yelling renewed. "Oh! Don't step there!"

"Where're you going? What are you trying to do?"

"Look O-O-O-U-O-O-O. . . ."

Another rib came tumbling down from its matrix pedestal. Part of a cervical vertebra collapsed. The man could hardly have stumbled over more bones if he had been trying for a record. He arrived at the quarry's edge shame-faced and repentant, but his mumbled apology did nothing to repair the damage. In the midst of it, he interrupted himself, "Well! How was I to know?"

We had no ready answer, but we prepared one: a railing to ward off further depredations.

"If that won't stop 'em," Carl said, "we'll post a man with a gun."

So far, we hadn't uncovered a skull. With the half-dozen or more dinosaurs now represented, it was strange that we hadn't found a single skull. Before giving up my excavation work, I had laid bare in one spot a few fragments that Carl later pronounced skull elements. A similar lot, with a number of long, pencil-like teeth, had come to light under the awls of Thrapp and Lewis, but we had yet to come upon a complete mass of bones with jaws, nostrils, cranium, and all that went with them. "Brown won't stop," Carl said, "until he finds at least one skull."

One morning Rainsford and Green came upon an interesting little set of tail-bones, lying in a loop like a carelessly coiled rope; even the terminal caudal vertebra was present. Stretched out straight, the whip-like tail would have been fifteen to twenty feet long. It was lighter, daintier, than the bones we had become accustomed to, but it was wholly parted from the fifty-foot creature to which it had belonged. Lewis got out his camera to take a picture of this record of the death agonies of a creature thrashing about, lashing and twisting in this remains of an ancient swamp.

In a bit Green, working nearby, called to Brown, holding up a sizable slab of shale for inspection. "What have we here?" he asked.

Brown took a small magnifying glass from his pocket and examined the rugose surface. "Why, this is quite a find!" he exclaimed. "This can only be an impression of sauropod skin!"

Everyone gathered around, and the fragment was passed from hand to hand. The surface showed a series of small, rounded tubercles in an even pattern, such as shows on the surface of several modern lizards. Brown examined the spot from which the fragment had come, clearing away more of the area with his brush. The pattern extended for some distance along the course of several caudal vertebrae. What happened was plain; the creature's tail, before the flesh decomposed, had left a print in the mud, and the markings had remained as the mud turned to shale.

"This is the first sauropod skin impression ever known," Brown exclaimed. "We've found skin impressions of one of the smaller fellows before, the dinosaurs that roamed the solid land. But this is the first record of the big boys who stuck to the water."

Some of the best-preserved parts of the impression were pried out with care and carefully wrapped in rice paper. Brown took them to his tent for safe keeping.

Long shot showing the Howe Quarry, the suspended barrel for getting overhead photos, and the campsite of the museum team.

1934 group portrait of the Howe Quarry crew, showing Brown in center, Bird on the far left in the back row, Mr.

Howe himself at the far left in the front row, and, next to him, Dr. Rainsford of Rye, New York.

Bird and a plaster-encased sauropod femur (thighbone) at the Howe Quarry.

Howe Quarry in early stages.

A herd of sauropod dinosaurs (*Diplodocus*) confronted by *Allosaurus*, a carnivorous dinosaur. Note the whip-like tails of the sauropods. Copyright © by Gregory S. Paul.

Lilian and Barnum Brown at the Howe Quarry with dinosaur bones.

9

IT'S A GOOD THING 1934 came around only once in my lifetime; I couldn't have taken it twice. I began it as a wandering motorcycle cowboy, looking for almost anything to happen. And it had happened in spades . . . and awls and prospecting picks and dust brushes . . . on the most interesting job in the world. I was working with and for the top man in his field. And he was working on the biggest job of his life . . . fifteen dinosaurs, more or less, in one fantastic graveyard. And it had all happened because a fossil I had stumbled over happened to be the first of its kind ever found.

Added to an everything that needed no additions, the social side of the end of that one-of-a-kind summer outshone anything that happened along this line in my life.

The Patons had urged us repeatedly to rally around their ranch for a weekend party. They brought around Mrs. Ewen from the Ewen ranch on the flats below Shell, who also urged us to pay them a visit of a weekend. And the Austins, who had a remarkable collection of local fossils of many types gathered over the years.

We didn't start to cash in on these wonderful invitations until after Brown arrived . . . and after *The New York Times* told the people of northern Wyoming what a wonderful find they had, with a few words on what an outstanding aggregation of scientists had rallied here to help them develop it.

The social season opened with a party at the Frank Ewen ranch. It took two cars and a truck to get us there. After the ruggedness of the Howe place, the Ewen ranch was a million dollar movie set. It nestled in the center of a green carpet of alfalfa that sloped toward the snow-capped Big Horns. A row of cottonwoods graced the lawn.

"This is really an honor," Mrs. Ewen beamed.

"You know, Dr. Brown, we were almost afraid to ask you and your eminent colleagues down to such a simple home affair."

Looking over the assembled Ewens and our own sizable party, I was more worried about how they could ever fit us in.

Mrs. Ewen flitted graciously among her guests, chatting a minute here, a moment there, with Brown, Green, Rainsford, giving each his title, awarding Lewis his 'doctorate' a few months early, conferring professorships on Carl and even on me. "Let's soak it up," I said to Carl. "It's only for a night."

As dark came on, somehow everyone found a chair in the crowded dining room. After a hard day's "prospecting," we were all hungry enough to put a damper on conversation until the meal was done. Afterward, Carl and I helped the girls wash up dishes in the roomy ranch kitchen; Carl was sure our lack of erudition and our phony upgrading stood less chance of being exposed there, and I felt the company of the girls a nice change from camp.

By the time we were done, the party had wandered out to the pleasant porch, into the cool summer air. The porch lay in shadowy darkness, with dim backlighting from the lamps in the house. The thinnest of white light bounced from the glittering snows atop the Big Horns, sparkling down on us from the diamond-pointed stars. There was a bit of cricket music from the hedge, the soft scrape and shuffle and bustle of getting comfortably seated in comfortable chairs, and for a bit the pleasant insistence of Mrs. Ewen attempting to get Brown opened up for story-telling. Since we first had met, I had learned a good bit about this remarkable man, but when Mrs. Ewen got him going and he got his pipe properly stoked, I was amazed by a new and until now unrevealed facet. What a raconteur!

No one measured the time we spent, following Brown in his long fossil-hunting trail around the world. Sometimes he stepped outside the realm of fossildom; there was a mysterious luminous giant spider that got away, in India. He took us through a shipwreck off Tierra del Fuego, a camel trip across the desert in Ethiopia, a trip through Greece for fossil mammals . . . Crete . . . the Mediterranean. Now and then a puff on his pipe lit up his face in the darkness. We finally stopped with a lion hunt in Patagonia.

Brown was inside the cave, after the lion that had been raiding a local herdsman's little flock. Supposedly, the lioness had cubs within the confines of the cave, and Brown was advancing stealthily, gun in hand. Suddenly the lioness appeared, but not in front. Behind him!

Everyone gasped and twittered with excitement. "What in the world . . ." came from all sides.

Brown puffed again, and ran a smooth hand over his forehead. He had a notably high forehead, smooth and shining, running all the way across the top of his head and well down his neck. "That's when I lost my hair," he concluded.

The following week we partied at the Austins'. This was followed by an end-of-season party at Patons'. Shortly after the Paton party, Mrs. Brown joined our camp . . . a sparkling young woman with brown hair, great hazel eyes, and irrepressible humor. "What's Brownie been up to, before I came?" she inquired, and raised an admonitory hand as I opened my mouth for whatever inane reply. "Now, now, R.T. . . . don't tell me more than I want to know." Mrs. Brown fitted into the picture wondrously; we wondered what we had done before she came.

To enhance the party atmosphere, Paramount News sent in their photographers for newsreel coverage, and the old quarry took on something of a Hollywood air. Rehearsals were arranged, dramatic expositions of our work staged. Brown went through the motions of bone-cleaning in the quarry, while the rest of us rallied around as extras, bit players in supporting roles.

On the heels of Paramount News, the George Sheas from Billings, old friends of the Browns, joined us for a spell. I had heard of Shea before; he had once dropped food to Brown's party by plane when they had been marooned by an unseasonable snow. And the Sheas were joined by Trubee Davidson, president of the American Museum, who brought along his wife and son. Then the John C.

Germans, staff artists from the museum, dropped in for a short stay and a bit of sketching.

Most of this high society was, however, a bit outside my personal periphery. My bone chart was taking form, but it was taking tremendous pains and a lot of time. In terms of knowledge, I had started from near zero. I was beginning to master a few principles that were helping me catch up in the work; in the case of the larger bones, I could now fairly well project them onto my chart as soon as one end was exposed, fairly well estimating the location of the still buried portion.

I used to slip away from camp after supper and climb a little knoll above the quarry, the better to absorb a sight I was forced to look at all day but had little time to dream about. One evening I heard footsteps, and saw Brown climbing up to the same vantage point. He seemed surprised to find me there but grinned and moved over to where I sat.

"You won't see many more sights like this in a lifetime, R.T.," he said. Thereafter we often met at this place and time, to sit inspired by the dramatic story written in the bones. By this time parts of at least fifteen dinosaurs had been disclosed. Again and again we went over the sometimes questionable skeletal hookups. We had finally found a complete skull, undamaged and at the end of a long neck. The completely articulated tail with its curious loops was a prize find. A new big skeleton in the back of the quarry, still largely buried under overburden, still lured us on. We would sit chatting, weighing, considering, until the last light left the sky and the bones sank into the night, and then turn back to camp and the lanterns burning in the cook tent.

Somehow, I came upon a time when I had managed to push my chart work ahead a bit and returned happily to prospecting for and digging out bones. I chose a spot where the shale was firm. It worked beautifully and freed from the bone to perfection. An ilium and a pubis, part of a sauropod's hip structure, came to light under my happy hands. The hind legs stretched away to one side, and the bottom of the creature's abdominal basket lay under hand. Between the pelvis and the abdominal ribs, my awl rasped against something different, something that wasn't bone. I dug a little deeper, and a smooth round stone appeared among the fragments of the matrix. Close by another . . . and another. When I was done with this pocket of smoothed, rounded stones, they totaled sixty-four, all lying next to the abdominal ribs, embedded in the very spot where the digestive tract had lain. Scattered through the shale with them was an amount of carbonaceous material . . . old stems, bits of leaves, other plant matter . . all black or rusty red. Could this have been part of the old dinosaur's last meal? Here, in this long-ago shrivelling puddle where giants had trampled each other into the mud, had this old fellow held together better than the rest? I gloated for a little while and then went hunting for Brown.

"I've found some gizzard stones," I said.

He was as delighted as I. "R.T., this is a great find!" he exclaimed. "You know, R.T., there are geologists who don't believe in dinosaur gizzard stones. But when you find them still in the old boy's belly . . ."

August—an August blazing hot for this high country—and the work still went on, Brown driving ahead with undiminished interest in each day's finds. But Carl was growing perturbed. "If we don't start taking up bones soon," he said, "we'll never finish the job this year." As the quarry expanded, the problem of keeping exposed bones protected from bad weather grew; a heavy rain would bring tremendous damage to our specimens, lying as they were each on its pedestal of clay or shale. They had been generously doused and redoused with shellac, of course, and would individually hold up fairly well, but the dried-out matrix, loose and porous, was developing a tendency to crumble when wet.

We had, early in the season, borrowed a haystack tarpaulin from the Ewens, a canvas about twenty-five feet long by twenty wide. As the quarry had grown, we had added several of these. Excellent protection, but with one drawback: it was ticklish business to draw these heavy tarps over the bones without knocking over the ones that projected above the general mass. Men lined around all sides could carry a canvas and lower it over a portion of the quarry, but those grasping the forward edge had to find a way across the exposed bones. To achieve this, we had placed sandbags here and there to step on. Fortunately the pitch of the beds provided a fair drainage for the tarps. "Will it rain tonight?" was invariably asked at the supper table. When it had been thoroughly hashed over, Brown would issue the final decision to cover or not to cover our boneyard.

One night near the season's end I was awakened around midnight by the hot silence. And thick darkness. Then the world outside suddenly flared into incandescence. Instantly with this there was a sharp crack, followed by a heavy roll of thunder. I sat up, frozen in the bedcovers; we had guessed wrong about the night's weather. I struck a match, trying to warm up the generator of my gasoline lantern with trembling fingers, while the thunder outside bumped and thumped and rumbled down the wall of the sky. Outside, sounds of frantic motion came from the nearby tents.

Lewis called to Carl, "Wake Frutchey. Just dump his cot!"

My lantern caught, and I was soon outside. Outside other lanterns were bobbing up the hill toward the quarry. Brown hurried by, a coat wrapped over his pajamas. Another flash of lightning seared the darkness, as I found myself stumbling up the path on Rainsford's heels. In the quarry lanterns glowed and bobbed fitfully among the great jumble of dinosaur skeletons. We gathered up one of the tarpaulins and unrolled it in an open space beside the path, stretching it into a great taut square of cloth, and began to move toward a sandbag crossing of the quarry. I heard Brown groaning a few feet away, "God's sake, boys, don't step on the bones!"

Another flash of lightning. The thunder that followed was louder than before, and mingled with it was the low moan of the rising wind. We felt our way slowly among the bones, guided by flash-lightning. A brief stir of air, and then the wind struck, a mini-tornado. It caught our white burden, turned it into a viciously slapping sail. One of my fumbling feet missed a sandbag, fell on something that gave way with a crunch. Over the wind's sound I could feel rather than hear the sound of crackers breaking. The canvas beat up and down on the bones, and again I heard Brown moaning, "Watch out for the bones, boys! Oh, God, the bones! The bones!"

We tried to bring down the slatting, slapping cloth; we forced it to the ground; we got one edge down and weighted it with rocks. Then the rain came . . . big, cold, hard-slapping drops. They pounded on my bare arms and through my thin shirt; they wet Brown's coat and his pajamas . . . and they stopped the flapping of our big sheet. We moved quickly to unfurl the next tarp. Again we struggled across the quarry in the noisy, flash-lit dark, but now we had the help of the beating rain. We covered the still-exposed odds and ends with smaller tarps and headed back for the tents, as wet as if we had fallen into Shell Creek.

Next morning we dreaded to lift the canvas, fearful of the casualties we might find. We found them. But the quarry had been spared a bad wetting. We patched the damaged bones as best we could, thankful things turned out no worse.

In fact, the storm may have done us a good turn. We were in high country, a world whose residents joke that they have but three seasons—July, August, and winter—and August was already more than half gone. The storm served to bring Brown, the tireless giant-hunter, face to face with the season. We knew now the job had no end, anyway. We had reached the north and south limits of the giants' burial ground, but the center went on indefinitely under the bank. We now had to move six cubic feet of sandstone to get at one square foot of bone-bearing shale. So the day came.

"We'll start taking up skeletons soon's we finish clearing shale from what's already uncovered."

Milo Howe's team was retired to the barnyard, and Milo himself, as well as his helper, were added to our crew of excavators. When I caught up with charting, I went to work behind the big skeleton at the back of the quarry where the tail curled toward the bank. I worked along the caudals, hoping they would curve again before reaching the heavy sandstone. No luck; I reached the last two bones that could be uncovered without hauling away more sandstone.

Brown stopped by. "R.T.," he chided, "why did you let his tail get away? You know when we'll be done here if we have to chase him under that rock?"

About this time, *Diplodocus* landed at Greybull. *Diplodocus* was a four-passenger cabin plane Brown had hired for the next two months. It was piloted by Harry McIntyre, and on its side was painted a picture of the dinosaur giant for whom it was named. With this Brown planned to survey all the Mesozoic outcrops, the three dinosaur ages that occur in the west from Alberta to Mexico. This was a Sinclair Oil & Refining Company project, but at the same time Brown would have an eye out for new areas for hunting dinosaurs in the years ahead. Before Brown left on this survey, we were flown over the area to give us some idea of what sort of country we had been working on. The Howe ranch, from the air, was shockingly tiny, for all the work we had put into it, and we couldn't even see the bones on which we had grubbed for so

long. Red Chugwater was a toothed saw against the grey Permian limestone. Jurassic Sundance lay below it like a twisted pink ribbon. With the overlying Morrison, it filled the space between the Triassic and the brilliant colored beds of Cloverly . . . all Dinosaur Time neatly and colorfully stacked between its stone boundaries, a truly colorful sight. "From the air," Brown pointed out, "you get an overall concept of the formations and their relationships that you can't get from the ground."

The next week, taking Lewis along as photographer, Brown left on his survey. Happily for our little society, he left Mrs. Brown with us. And just before leaving, he hired two new men to help with the bone-packing: Dallas Hurst, an unemployed sheepherder, and Ben Allen, a nearby rancher who had a bit of slack time on his hands. I was now fully employed with charting, blocking in the bones that were under the bones we were removing.

We still had visitors . . . in fact, the crowds were getting larger as time was pressing down on us. We all enjoyed attention, of course. And good public relations are important to museums. But at times it was tiring to push on with our work and still avoid stepping on innocent and interested bystanders. Our newest man, Dallas Hurst, had the right idea in a way, but we "museum people" felt we couldn't copy it.

"Look at Dallas," someone pointed out. The stocky little herder was going about his new job with vigor and dispatch. In the front of his hatband was a hand-lettered sign, "No questions answered." A sign in back was more firm: "Keep off the bones!!"

I was just drawing in the limb bones of a little dinosaur on my chart when I met for the first time a gracious and grand old girl, the sort anyone would love as an aunt. I met her sisterhood many times in the years ahead, but we always remember first times. I was, at the time, being driven . . . driven by the calendar and by a job that had outgrown us. But I was taught to be nice to old ladies, no matter what.

She leaned over and touched my shoulder gently when I raised my pencil from the paper. "Young man," she asked, "how did you know where to start digging?"

I explained that certain bones had been found weathering out of the hillside. I pointed out the six-foot layer of sandstone that still remained at the back end of the quarry.

"That sandstone," I explained, "once covered

this whole area. It had to be cut down and removed to get to the bones at the back of the quarry . . . to get at the dinosaurs lying beneath it."

I stood up and warmed up to my story; she was, after all, a Grand Old Girl and she was attentive. I repainted the picture Brown had painted for us. A dying, shrinking lake. These great stupid behemoths retreating, dropping back to the shoreline of this mud-puddle, dying, finally . . . dying as they stood on the backs of those who had died before them. The bleak hot horror of it almost choked me up. Before I could unchoke, she raised a polite hand.

"Not so fast, young man," she begged, "please.

Will you just go over that last part again, please?"

I replayed the last rites of an ancient tragedy, and she listened with the raptness and care she had given me the first time. She looked again up at the sandstone layer, down into our bone pit, and back at me quizzically.

"Well," she said, "all you tell me may be so . . . if you say so I guess I have to believe it. But I still can't see why such creatures would have wanted to do that in the first place."

"Do what, ma'am?"

"Why, crawl away back under all that rock to die."

Gastroliths (stomach stones) *in situ* among the bones of a sauropod dinosaur, Howe Quarry. Photo by Bird. Bird reported that the gastroliths were found "between the pelvis and the abdominal ribs . . . all lying next to the abdominal ribs." The large bone below the brush appears to be the left ilium of the pelvis. Projecting downward and to the left of the main mass of the ilium, underlying a metal probe and forming the lower boundary of the space containing the gastroliths, is a process of the ilium which would have articulated with the pubic bone of the pelvis. Just to the left of this process is an elongated bone running from the bottom of the photo toward its top, probably an abdominal rib.

IO

CARL PICKED UP a bundle of empty burlap bags, handed them to me at the entrance to his tent, and went inside, to reappear in a moment with a ream of rice paper, a pair of sheep shears, and three pairs of rubber gloves.

"When we get these sacks cut into bandages," he said, "and when Thrapp gets back from the spring with a couple of pails of water, we can start plastering bones."

We climbed up to the quarry. Brown had flown back from Montana and gone on again, leaving Lewis to assist in the job ahead. However, our personnel had been cut by the departure of Rainsford, his son Laurence, and Green, whose summer vacations had come to an end. The final excavation into the sandstone had been made, revealing the last of the big tail and many other bones but showing no end to the bone layer and turning up no new skulls. The skeletons had been photographed from all angles and from the top of Milo Howe's hay derrick set up to give us an "aerial" view. Except for Ted Lewis, the rest of the crew were at work cutting a drainage ditch in back of the quarry to take care of possible heavy fall rains. The bones lay gleaming in their final coat of shellac. It was going to be a job to remove and package this tremendous mass of rocks for shipment.

"Where do we start?" I asked.

"Next question: 'When'll we be done?'" Carl said. "Makes little difference where we start, so long's we move in from the edge. You get this cervical vertebra, over here by itself, down on your chart?"

"I sho' have."

"Well, then, let's start right here. But first we'll have to cut bandages."

Carl sat down on a folded tarp, picked up one of the empty sacks, and began to unravel the seam along the edge. I followed his example. When we had opened several, he stacked them neatly on the ground, picked up the sheep shears, and cut them into strips four or five inches wide. These, in turn, were cut into twelve to fourteen inch lengths. Thrapp appeared with his pails of water and opened a sack of plaster of Paris that stood in the back of the quarry. Nearby was a pile of short cedar sticks, gathered from the growth in a draw back of the ranch house. Lewis was starting to wrap the many small bones in newspapers . . . claws and toe bones and the like that would not require a plaster jacket.

Carl and I moved over to the cervical vertebra. The complex mass of spines and processes with a rib prominent was about twenty-eight inches long, looking as if it might weigh thirty or forty pounds. It was endlessly fractured, but had been soaked and soaked again with thin shellac and was quite hard and firm. Carl cleared away the loose shale around it with his pick, undercutting it until it stood on a pillar of matrix several inches high. Then he brought over one of the buckets of water with the ream of rice paper and a whisk broom. He passed me the paper.

"Now if you'll just hand me sheets of this as I ask for them . . ."

Carl dipped his whisk broom into the water bucket and with quick flicks of his wrist sprinkled the surface of the vertebra. He reached for the first sheet of paper and laid it across the end of the bone, wetting it further with additional flicks of the broom. The water drops beat the fragile paper to the bone surface, where it adhered evenly. Carl patted it down and said, "This paper keeps the plaster from sticking to the bone." It took two more sheets to cover the exposed surface.

Meanwhile, Thrapp had finished cutting bandages and was mixing plaster. Dan appeared at my side with a pail of the white mixture. Carl put on a pair of rubber gloves—working barehand in plaster of Paris is worse on hand texture than protracted dishwashing. He reached for one of the bandages and dipped it into the milky mixture. "One thing you must watch out for in plastering bones: don't get your plaster too thick. Dan's got it just about right."

Though the mixture seemed to me as thin as water, there was plenty of white color on the saturated burlap; it looked as if it had been dipped in buttermilk. Carl laid it on the bone, lapping one end beneath the undercut edge, the other across the top. The wet burlap clung to the dampened surface, helped a bit by a few taps from Carl's fingertips. Thrapp slipped on another pair of rubber gloves, dipped another burlap strip, handed it to Carl, who laid it in place with a generous overlap. The process was repeated until the stone neckbone was covered with a double layer of plastered burlap.

Carl walked over to the pile of cedar sticks gathered from the draw back of camp, selected one about the length of the neckbones we were working on, and laid it alongside the piece. There was just enough plaster left in the pan to moisten the strips of burlap enough to bind the stick in place. It made a fine splint, a stout support for the casing.

"There," he said. "That's all there is to it . . . except to turn the block when the plaster sets, and plaster the other side."

A visitor on the sidelines spoke up. "Do you expect to do that to all those bones?"

"Each and every one."

Carl moved on to the next stone bone, this time allowing me to help. And on to the next. In an hour the first specimen had set nicely and was ready for the next step. I referred to the quarry chart, on which each square had an identifying number. With black paint I numbered the bone jacket to correspond with its square on the chart. Carl arrived with his pick, and we soon finished undermining the plastered neckbone. The burlap casing was now quite rigid but still damp, and Carl treated it with respect as he lowered it from the pillar to the ground, gently turning it upside down. It took only moments to strip away the remaining matrix with our awls, and the newly exposed surface was treated with shellac.

"Now if this hardens before evening," Carl said, "we can apply the rest of the jacket. Otherwise, we'll catch it first thing in the morning."

By the end of the week I felt almost as skilled as Carl in slapping plaster jackets on old bones, but of course there was no real problem in these bones along the edge of the quarry. It was a highly repetitious job . . . fascinating boredom.

September hurried by. Late in the month Frutchey left for the East, and Lewis nailed his discarded ski boots to the boards that framed the doorway to the cook tent. He was replaced by Mrs. Denniger, brought in from Montana, who had cooked for Brown on previous expeditions. A stocky, good-natured woman who soon became as much a part of the cook tent as her pots and pans, she was known to all of us as "Maw." "If the way to a man's heart is through his stomach," she told us, "you boys are gonna love me."

And we did.

II

OCTOBER BROUGHT with it warnings of impending winter. An unexpected snow squall near September's end had disrupted quarry work for a week and left some of us, including me, marooned at the Patons, where the heavy fall of soggy flakes caught us on the way back from Greybull. The delay threw us further off schedule, but Indian summer came in with the departure of the snow and gave us a welcome reprieve. Nights were chill, but the golden days, neither too hot nor too cold, were a delight.

But only the weather was delightful; on all other fronts, we had a mounting feeling of pressure, of hurry, of inadequacy. Ted Lewis had to go back to Yale. Mrs. Brown, who had been the Queen of the Camp while Barnum was away, left us for California. Brown was still engaged in his aerial survey for Sinclair, repeatedly advising us when he would be back, repeatedly extending his stay away. We put on another plasterer, Red Snyder, but he couldn't be expected to carry the load of the experienced people who had left. And from time to time, even with the unexpected and extended Indian summer, we were daily reminded by one or another of the area's natives, "Here we got only three seasons—July, August and winter." The days stayed unexpectedly, even unreasonably, dandy. But each day was predictably just a few minutes shorter than the day before.

Bit by bit, too, the job was getting in the way. The back end of the quarry was daily piled higher and higher with plastered bones, while inside the quarry the spaces left by the ready-to-go fossils seemed insignificant in comparison with the tremendous job still ahead.

Naturally, we had plastered and moved the easy, uncluttered fossils first. Near the back of the quarry, more and more in the way of our plastered bone pile, was a great mass I had worked on long and randomly, extricating the components from the matrix and the mass. The mass was dominated by a perfect five-foot scapula that lay partly over and partly under a tangle of ribs, some assorted vertebrae, and a long section of tail. Days had been consumed in untangling from one end of the pile the great ribs that lay across it like broken barrel hoops. I had isolated and plastered the loose vertebrae, and now the shining black scapula was completely exposed, but resting on still other ribs in such a way that there was no way to properly underlap the bandages of the jacket. I worked with special care, bracing bone with stout pieces of cedar incorporated into the plaster. When it hardened I called on Carl for help in turning the bone, warning him about its uncertain condition.

"It looks bad, all right," he said. "but all we can do is hope and grunt and.pray. We'll just have to lift it and hope it stays in its jacket."

We freed the great scapula by forcing our steel digger blades between it and the tail bones. It creaked and cracked as it came loose from its resting place, and there was just room enough to stick in our fingers and grasp it from below. We lifted and started to turn the great, beautiful blade, its shiny black accentuated by its plaster pillow, doubly important for being part of a coherent group. We couldn't . . . we didn't dare . . . let it come to harm.

Just as we had it almost down, there was a dull clatter and our burden lightened. We heaved what was left of our burden onto the sandbag I had readied for it, but we were standing now in a clutter of black fragments. Carefully we gathered them up, bagged them, marked them properly, and tucked them in with what was left of the beautiful black bone. Since it was all part of one of the most complete skeletons we had turned up, we couldn't sweep it under the rug. Bit by bit, it must be pieced together back in the museum laboratory.

"Can't win 'em all," Carl said as we plastered the broken package shut. "I just hope I'm not around when it's opened. Dull job!"

Added to the pressures of such mini-catastrophes, the inexorable sweep of the season, the shortening of the daylight hours, the awesome size of our find, the important but repetitious task of answering tourist questions, there was the pressure of another sort of "fossil degrading." This grew on us in direct proportion to the daily growth of the tourist crowd.

The questions were the worst in a way, because they were so predictable, so repetitious. But it was important to preserve a pleasant image of scientists, diggings and museums with a smile or an inwardly hollow chuckle.

"When do you folks expect to finish?"

"November first if we keep right at it and everything goes well."

"Which November? You mean this year?"

The sightseer would let his eyes ramble across the bone field, dwell briefly on the puny pile we had treated, drift back to us.

"You mean this whole quarry or just what you are working on?" Turning to a fellow-tourist, "I can't believe it. Doesn't seem possible; does it, Joe?"

"Why, if that little man over there with the plaster on his face wraps up five more bones by Christmas, I'll eat his pail, plaster and all."

As we became more hurried and pressured we were less able to be observant about anything but our stone bones and their care. A portly gentleman I noted one afternoon will do to illustrate. Perhaps I was less busy than usual, or perhaps he was unduly nonchalant. He squatted near the upper margin of the quarry and ran manicured fingers gently along the contour of a rib, seemingly concerned with its texture.

Moments later he stood up, slipping his hands into his pockets, and sauntered down the path leading by me and out of the quarry. I stood up and looked at the rib where his hand had rested. It was now a shortrib. The man paused, took an interest in another bone, called attention of another visitor to it, talked about it a moment, started on to squeeze by me. I stepped forward and held out my hand.

"Sorry, mister, but we need all the pieces we can find."

"Wh–what do you mean?" he asked innocently.

"Well, it's this way: when we put all this together back at the Museum, we need all the parts we can get. Even then, we have it rough enough. Would you believe that of the seven tyrannosaurs known in the world, no one of them is complete? We've had to borrow around from each other to paint a full picture. But we try all we can to keep this down to a minimum. So. . ."

"So?"

"We have no idea until we get back and start putting this all together how important that piece in your pocket might be."

"That what?"

"You see that short rib over there, where you were sitting? We won't know for months how important that piece in your pocket might be in the over-all story of that dinosaur."

The shame-faced bone-borrower surrendered the fragment and hurried on. He didn't saunter.

I would not include this incident but for the fact it was not isolated, nor was it the last of its kind. The next day several claws disappeared from one of the articulated feet, where all the bones had been laid out neatly by the dinosaur's Grim Reaper. I can think of no finer souvenir than a dino claw; I can understand the purloiner. I just can't string along with him. The dinosaur bone business is always in short supply on foot bones and skulls.

But to get on with it: the little man with the plaster on his face met November first with literally hundreds of bare bones still waiting for cover. Bit by bit, colder mornings and shorter days took much of the fun out of what had been a pleasurable task. Plastering settled down to a cold, hard, boring grind. The sullen skies warned us every day of what they might turn loose on us without notice.

At the lowest ebb, Brown came back, giving us all a needed shot in the arm. Carl and I picked him up at the Greybull airport. He dismissed *Diplodocus*, the survey plane, and rejoined the expedition. This changed the pace, though until he arrived we felt we had been driving ourselves manfully. We were all aware that August was away behind us and that western winter might move in any day.

All of us who knew Brown were prepared to see him pull on his work clothes and join the plasterers. Riding high for two months hadn't spoiled him, and after all, keeping Harry Sinclair in our corner had been as important as our work here in the boneyard. More important; Sinclair fed us. The aerial survey, in which Brown was the pioneer, paid off in bringing attention to oil prospect areas that earthbound geologists, too close to the trees to see the woods, couldn't see nearly so well.

Milo Howe went to work building boxes for the mounting pile of wrapped and plastered bones. Ben Allen and Dallas Hurst, if we were to end the season as winners, had also to be detached from the plaster job and set to boxing the prepared bones, each tucked carefully in a bed of golden straw. We weren't surprised when one evening Barnum suggested we omit covering the field with the usual tarps at quitting time.

"I'm coming back after we eat," he said, "and get in a few extra licks by lantern light."

We went down to the cook tent in the gloomy dusk. "Maw" had supper on the table. After a day in the chill, wet plaster and the shivery air, it felt good to sit around the warm fire. But with no further announcement, Brown pulled on his sheepskin and picked up a gasoline lantern.

"Nobody else has to do this," he said. "And after all, *I* was out flying around while you fellows were all grubbing. But it looks like night work or no finish for this job this year."

None of us hung back . . . maybe dragged our feet a little. We gathered up a pile of cedar brush and touched a match to it. The brushy wood crackled and snapped. Flames flared up into the black night sky. The remaining skeletons were no longer mere massive bones on pedestals of clay; they became shapeless patches of white scattered through lumps of darkness; they caught the wavering firelight, twitching and quivering like phantoms that might at any moment rise and fly off into the night. The fire was warming enough at close range, but the fearsome white shadows were even colder and more unpleasant to work on in the dark.

I set my lantern down a few feet from a five-foot block of dorsals and ribs, gave a few experimental digs with the prospecting pick to undermine the plastered specimen. The work was doubly difficult because the pick and my slowly congealing body both cast shadows into the wrong places. My fingers were freezing inside the clammy rubber gloves. At the fire I tried to thaw them. A dull ache crept up my arms equally from the cold and from the sharp pain that came of warming over-cold hands before an over-hot fire. Alternately, we all shuffled up to the fire, back to the job. Back and forth.

We stood it for two hours. Even for the insatiable Brown, dinosaur-hunting may have palled a bit.

12

THE NEW YORK TIMES and the *Herald Tribune* both carried stories of the returned expedition. It was, after all, more than a run-of-the-mill thing, even for the American Museum of Natural History. Brown had set out on the trail of a simple, normal dinosaur or two and had stumbled into a graveyard of dinosauria so vast, so full, that it was overwhelming. Pictures properly adorned the story. Barnum Brown standing beside a great box of bones in the museum's basement. Me standing right next to him, supporting a huge femur taken from the back of the quarry. Behind us, rank on rank, sixty-seven thousand pounds of plastered and packaged stone bones. It still irks me I was introduced to readers by a careless reporter's accident as "Robert Burns" rather than as Roland Bird, but they got Barnum identified properly; he had been in the New York papers before.

Overriding the gall of the error in my name was the great news Barnum had for me: a staff member of the Department of Vertebrate Paleontology was retiring at the end of the year, and Barnum had secured his place for me; I was now a proper and full-fledged fledgling of the staff. From my earliest years, the memories of that great red granite building on Central Park were sharper and more all-engulfing than memories of my cradle and high chair. The brass signs of the zodiac inlaid in the tiled floor of the entrance, the great thirty-six-ton meteorite, the tremendous model whale suspended from an inside ceiling, the *Brontosaurus* and his smaller companions on the fourth floor . . . to me, all this had been there when time began and would be there at its end. I couldn't have dreamed, there on Father's shoulders, that one day I might become part of the scene and that near the end of my days I might work to enhance and improve the *Brontosaurus*.

On the first day back from the field I went to revisit the laboratory, where I had first met Barnum Brown. The long series of big rooms on the fifth floor still fascinated me. The strange heavy odors still hung in the air, but the noise of Otto Falkenbach's hammer was absent. Otto was putting the finishing touches on an armored dinosaur resembling an oversized horned toad. I met Bill Thomson, who had been in the field collecting when I had been here a year ago. In the next room, Charlie Lang, Barnum's artist in iron, welcomed me to the laboratory force.

Compiling and consolidating the quarry chart, which had been my baby from its beginnings on the Howe Quarry, was to be my first job. The scale I had adopted, an inch to a foot, necessitated using a huge sheet of paper or Bristol board at least seventy inches long by sixty wide. I managed to get such a sheet from museum supplies, but when I looked for a place to spread it out, together with the sketch sheets drawn at the quarry, there seemed no suitable place in the whole department. I ended up taking the whole thing home to Rye, where I spread it out on the parlor floor. I estimated the job at a week to ten days. By the time I had transposed some three thousand bones from my field sketches to the master sheet, Father and the family had bypassed the parlor for a whole month.

Brown was delighted with the chart. With some difficulty I managed to unroll it full length on the table in his office. Never before had such an aggregation of dinosaur bones been drafted in such detail.

"R.T.," he was good enough to tell me, "this is surpassed in value only by the bones themselves." We discussed various hook-ups, trying to decide which skeleton to work on first. "You should start on something easy," he told me.

He led the way across the hall into the lab and to Otto Falkenbach's table. "Otto," Brown said, "R.T. is going to start in working on the Howe Quarry collection. Can you find room here to give him a table and maybe show him a bit about preparing bones?" Otto began to clear away bottles, brushes and bric-a-brac from a square table. "What are you going to tackle first?" he inquired.

Brown looked at me. "How would you like to try the looped tail?" he asked. "It's simple, but it will make a good exhibit."

This agreed upon, Brown suggested I take the rest of the day off and find quarters; as yet, I had no place to stay in the city. I was lucky to find an old brownstone not more than ten minutes from the museum. Next morning Carl helped me to get the boxes containing the looped tail upstairs. Otto told me Brown planned to have the vertebrae transposed, just as they had appeared in the field, to a bed of plaster colored to represent matrix. The tail, its graceful curves and turns down to the terminal tailbone, had been taken up in three blocks. We laid all three of these on separate tables, careful to place them all with the same side up. Otto spread newspapers over the rough field jackets of burlap.

"First, we have to give them plaster beds," he said.

He had me mix a tubful of plaster of Paris. When it started to set, thick enough to ladle out without running, we covered the top of each specimen with a mass of gunk about three inches thick. This settled into all the irregularities on the surface below the protective layer of newspapers but didn't penetrate the flexible covering. To give the plaster added rigidity, we thrust pieces of heavy wire into it while it was still plastic. In a few minutes the plaster hardened, and we turned the piece over.

"Now," Otto pointed out, "you've got the bones on a foundation comparable to the clay bed you found them in."

With a wet sponge he moistened the upper sides of the old plaster jackets. The water soaked into the dry, porous surface. Soon it was possible to strip away the burlap bandages one by one. They lifted easily from the bones, bringing with them the no-longer-needed rice paper. Wherever a bandage looped underneath, between the bone and the prepared bed, we cut away the free portions with a knife.

Willie Booth, the department's janitor, appeared with a broom. He was a small man, slightly bent, with sparse hair and few teeth. He looked down at the rubble on the floor, removed a pipe with gnarled fingers.

"I hope you keep the place cleaner'n Otto," he said. "Otto always leaves a mess."

"Now, Willie . . ." Otto laughed, as he went on stripping bandages. Following this, all bone surfaces were treated to another coat of shellac. While this dried, we went ahead with stripping away any spare matrix that had been included with the bone where fragile spines and chevrons had needed support or protection. The job took the rest of the day and part of the next morning. Then Otto again covered the blocks with loose sheets of newspaper.

"We'll have to have new beds to turn them on," he said.

The beds were prepared as before. They hardened and we turned the blocks, discarding the first plaster bases we had made. What remained of the plaster jackets was moistened with water and lifted free of the bones. The bones were cleaned up and daubed again with thin shellac. When this was dry, it was possible to lift each vertebra from its cavity in the plaster bed. This enabled us to cement any loose pieces together. The relationship of the bones to each other had, of course, not been disturbed.

"The next trick," Otto said, "will be to prepare a final bed, with the tail assembled as one unit."

We laid the three blocks together on a large table, each placed in its original relationship to its mates. Otto dumped a box of loose sand beside them, bedding them in it, piling the sand slightly higher than the edge of the blocks.

Now Otto applied a thin coating of lard oil to the shellacked surfaces of the bones and to the faces of the plaster beds on which they rested as well. Next, a large pan of plaster was prepared, tinted the color of the matrix, and ladled over the block.

Because the oil acted as a separator, when this last bed was set and dry it was easily lifted free of the bones. All around the edge, where the sand had been piled, was a roughness resembling freshly quarried rock. We placed this on another table and, when the last moisture had evaporated, transferred each vertebra to its proper depression. With a little recleaning, the long looped tail, black and shining against its grey plaster bed, appeared exactly as it had first been exposed in the field. The next day it was placed in the museum's foyer as a temporary exhibit, together with the gizzard stones I had found "in the belly of the beast," with a few pieces of shale showing sauropod skin impressions.

The chart had now been thoroughly studied by Brown. He decided we should work up one of the more complete skeletons, which had been taken from the back of the quarry. Some of the big dinosaur's bones presented problems not encountered in cleaning and mounting the curiously looped tail. But I had been mastering museum lab practices, under the tutelage of Brown, Carl and Otto, and I had come to think by spring that there was nothing I couldn't do with material taken from Howe Quarry.

One day Carl and I brought up from the basement a large box containing the pelvis of the creature on which I had worked at the quarry, by far the biggest job I had yet tackled. The box was opened, and the straw packing removed. The mass within was a single irregular block, weighing six or seven hundred pounds, far too heavy to lift to the table without using the traveling chain hoist. We spread some of the loose packing straw on the tiled floor beside the box and gently turned the pelvis out on it. Some of the burlap bandages looked weak, after the rough trip by freight from Wyoming, but for the most part they seemed sound and firm enough.

"Let's put a bed on one side and roll the specimen over onto it where it lies," I suggested. "Then we can get a chain around it and hoist it onto one of the small tables."

We prepared the bed. When the plaster had hardened, we called in Otto and Jerry to help with the mass. They came, flexing their muscles and looking to me for instructions.

"Which way do you want to turn it?"

I spread straw to cushion a spot directly under the chain hoist track. Otto looked at the block critically.

"Have you got it reinforced?" he asked.

"She's ready to roll."

We all caught hold under one side and put our muscles into the lift. The heavy block began to roll onto the straw, rotating with its bed toward the prepared cushion. It was half over. Then came a horrible sound of rending. Instantly the mass lightened just as it should have grown heavier. A rattling shower of bone fragments cascaded about our feet. There's something special about Howe Quarry bones: when they break up, they really break up.

The burlap jacket had busted open. We heaved mightily and in haste to avert complete disaster. With the block turned upright on the plaster bed, the shower stopped. The jacket lay like an empty chrysalis, a gaping split along its entire length. All too evident was the cause of the accident. A weakened bandage had given way where the most stress occurred. Others, unable to bear up under the added strain, had likewise parted. There still remained a portion of the pelvis, but judging by the size of the pile at our feet, more than half of it had fallen out. It wasn't the first tragedy of the sort I had seen, but it was the worst. And it was mine, all mine. We stood frozen at the sight of our mishandled handiwork.

Someone came up with a solution of sorts. "The first thing we better do is clean up this mess before Brown sees it; no need to make him sick too."

Carl and I brought up some empty trays from one of the nearby cabinets. Otto and Jerry raked up the yellow straw from the tangle of bone and matrix. Broken bits of the sacral vertebrae, pieces of an ilium and many small fragments of less easily identified bones lay mixed in the grey shale. Some of the larger pieces had obvious contacts with what bone remained in the jacket, and these we put back into place. The rest, we piled in trays and concealed in the cabinets.

Now only the ruptured jacket on its bed gave hint of what had befallen the pelvis. We placed this on a small platform truck with rollers, and I shoved it under a table in a dark corner. During the next few days I repaired some of the damage, working at times we were certain Brown wouldn't enter the lab. On occasions when he did come in, other material on which I was working served to give the appearance that all was well. When he finally asked about the pelvis, I told him we'd had a slight mishap in turning it.

I led him to the dark corner. I pulled the specimen out just far enough for him to peer under the table to see the edges of the rent. In the poor light he got only a glimpse of the horrible hole in the plaster casing, which I had arranged in the most favorable manner for his anticipated inspection. Brown looked at it critically.

"A good thing the whole pelvis didn't fall out . . . which it might easily have done," I pointed out.

Brown grumbled a few words of pained regret and left with the suggestion we do the best we could. He had no idea how well we had already anticipated this suggestion. The sacral vertebrae that had fallen out were pieced together, patched with plaster, and reinserted into position. The ilium was pieced and repaired. Only then did I have the piece hoisted onto the table to strip away the concealing bandages. At that, the damage was quite a shock to Brown. I resolved never again to unpack another block, no matter how small, without thoroughly acquainting myself with the condition of its bandages.

One day, while we were still buried deep in the Howe Quarry bones, Mrs. Lord laid a package beside Brown's pile of letters, a small thing wrapped in brown paper, addressed in a neat hand, no return address. When Brown unwrapped it and folded back the wrapping, a six-inch piece of black fossil rib fell out, a shiny black piece of rib, the Howe Quarry type. No message. Nothing.

"Somebody's conscience got to hurting," said Brown. Turning to me he added, "This goes to show, R.T.: you can't catch 'em all."

There was enough to do in the museum; sixty-seven thousand pounds of fresh fossils kept us from ever running out of ways to pass the time. Never-

theless, for the first time in years I was trapped for the summer in New York. Lab work was interesting enough, but I missed the field, the fun of digging for new and undiscovered things. Brown felt the same way but explained that funds for field work were lower than they had ever been. We should have to wait another year, he said, to test out the Rock Springs that looked interesting from the air when he was hunting oil prospects for Sinclair.

Perhaps the best thing that happened, taking everything into consideration, came of a suggestion of Brown's. "R.T.," he said as he paused by my table one day. "I think it might be to your advantage to attend Gregory's class in comparative anatomy; he's the best in the field, and you're short in formal background. Besides, as a staff member, you can attend free of charge, with time off for attending class."

So, for the first time in my life, I became a part of the college scene. Having quit school in the ninth grade, I had either missed a lot or escaped a great deal.

An overview of the Howe Quarry.

13

THE FALL of 1935 slipped quickly into winter. I enjoyed working in the warm lab, with snow swirling by the big windows just beyond my table. But by spring the yearning to get into the field again yanked at me. I had worked long and hard and faithfully on the Howe Quarry material, and all of it was interesting. But the fact it was all from my big adventure made the call of the wild even stronger. Otto Falkenbach had developed a few bones from the Howe Quarry, including a complete skull, but the rest had fallen on me. Brown had hoped to increase the lab force to speed up work on this collection, but, as with other things, we lacked the financial means to do so.

One morning on the way to the museum, just as the trees along Seventy-Seventh Street were tentatively testing their new season's buds, I ran into Brown. "It would be great to be getting ready for a trip back to Wyoming," I said to him wistfully.

Brown smiled. "I agree, R.T.," he said. "But it looks like another season without an expedition west."

The thought was depressing. Manhattan, glamorous and artificial, was pleasant if taken in small doses, but I rebelled against another summer like the last.

I had read of Triassic outcrops in New England. If I couldn't go west, maybe I could get to do dinosaur-hunting around here. Though I didn't talk this over with Brown, I thought a visit to the museum library might be in order. Taking out several books on local geology, I made the fascinating discovery that the Palisades, in New Jersey just across the Hudson, were old Triassic intrusions. The next Sunday I walked across George Washington Bridge to have a look at them up close. A dark grey basaltic rock, hard and dense, met my inquisitive eye. Here an ancient lava flow had forced its way into beds of sedimentary rocks and hardened.

I climbed down to the base of the cliff, where the Hudson ripples along a thin red outcrop of red sandstone, where tides rise and fall twice a day. The red sandstone, too, was Triassic, about two hundred million years old. The exposure was limited, but it represented muds laid down at the same time as those of the Painted Desert of Arizona. I followed south in excited search for bones or tracks; it was thrilling to reflect that here, almost in the shade of the Empire State Building, were rocks that might carry dinosaur remains. But they disappeared under the Fort Lee Ferry docks. I found nothing in the hard, sandy band of red sandstone, but I was out scouting again outside the great, wonderful, incredible, fascinating, confining stone barn of the American Museum.

Next day I investigated a partial skeleton stored in the museum, a skeleton collected from these very exposures in 1912, a phytosaur. The creature, resembling a crocodile, would have been some twelve to fourteen feet in length. He was neither a dinosaur nor a crocodile but possessed affinities to both. Nearly half the bones of the museum specimen were present and complete, prepared with the readily recognizable red sandstone matrix against which the pinkish white chalky bones showed up well. I looked it over carefully, knowing it would probably be typical of any bones found in the eastern Triassic.

Continuing my study of regional geology, I learned there was a considerable territory up in the Connecticut River Valley, extending all the way up to Turners Falls, Massachusetts, that was Triassic in age. Lull's *Triassic Life of the Connecticut Valley* revealed that dinosaur tracks and occasional scattered bones and parts of skeletons had long been known in places about the valley.

I went to see Brown in his office.

"Why can't I go up into Massachusetts and

Connecticut and do a little local fossil-hunting, since we can't go out west? Mostly, I can go by streetcar."

He looked pleased. "Who says you can't? Sounds good to me."

He spoke enthusiastically. "We could use a slab of trackway from Connecticut Triassic; it would look good on the ambulatory leading to Jurassic Hall."

The wall of which he spoke was part of the new Roosevelt Memorial Wing just completed. The flat white expanse was ideal to display great slabs of rock showing dinosaur trails.

"I've had this thing in mind for some time," Brown said. "After we pick up some New England Triassic tracks, others from other parts of the country can be placed along the wall as they are collected."

I first decided on a four-day weekend trip. With a feeling hard to describe, I packed a collecting bag with an awl, a couple of whisk brooms, a small hammer and chisel, and a clean shirt.

The train came into Holyoke, Massachusetts, in late afternoon, and I checked in at the Nonatuck Hotel. In early evening I walked across the bridge into the town of South Hadley on the other side of the river, wondering where I might find someone to tell me anything of local fossil spots. I tried the library across the street from the campus of Mount Holyoke College. A Miss Gaylord at the desk named several local people she knew to be interested in fossil finds thereabouts.

"But if you go to see Miss Cooley on Silver Street," she added, "she'll give you more information than I can. She's head of geology at the college."

I almost ran to hunt down Miss Cooley. She met me at the door of her little white cottage and invited me in. When I told her of my purpose in coming to Holyoke, she rattled off a list of people to see. As I started to replace the envelope in my pocket, on which I had hastily jotted down names and addresses, she added, "Then there's Miss Talburt; maybe you've heard of her. Back in 1910 she found a little dinosaur skeleton not far from Mount Tom."

I recalled reading about her in Lull's book. The specimen was a small bipedal carnivore, one of the smallest. Miss Talburt had found it in a block of sandstone in a nearby field, but that was all that was ever found in the locality.

"And then, there's Carlton Nash, up at Moodys Corners. Recently he made an unusual find. I haven't seen it yet, but from his description it's not like anything ever found around Holyoke." She paused, seeming at a loss for words.

"He says it's a place where a dinosaur sat down."

"Where a dinosaur what?"

She laughed. "I know it sounds fantastic, but Carlton is certain that's what it is. Of course, Carlton is inclined to be certain by nature."

I tried to picture a dinosaur sitting. Undoubtedly, the creatures must have sat, sometimes, even as cats and dogs. But sitting down in the middle of a Triassic mud flat?

Next morning I left the hotel with my prospecting bag. I was tempted first to visit the boy who had found the "sitting dinosaur" print, but he lived some distance out of town. The nearest prospect on my list was Donald Simpson on Franklyn Street. A good-looking boy in his late teens answered my knock. Mention of dinosaurs brought instant response.

"Sure, I know where there are plenty of tracks." He glanced over my shoulder at a car standing in the yard. "I'll run you out to one or two of the places, if you'd like me to."

Here was someone like the Patons, not only interested in fossils but willing and ready to share with anyone of like interests. In five minutes we were rolling out of town on a country road with little traffic. The region was slightly hilly. We passed open fields, a patch or two of woodland, an old stone wall bordering a gradual curve in the road. Donald drove off the pavement and stopped. Against a hillside, just visible across a grassy field on the other side of the stone fence, was an old stone quarry. Open pits and areas from which rock had been stripped extended several hundred yards along the base of the open hill. We climbed the wall and started for the quarry.

"We'd best go in this back way," Donald said, "and not bother Mr. Murray."

The remark seemed to require an explanation. "Who is Mr. Murray?"

"Oh, just the owner. He doesn't like people coming around, especially if they want to see his dinosaur tracks. But if he doesn't know we're here he can't object, huh?"

I had a feeling the situation might develop embarrassing angles, but we kept on.

"Do you often come here without Mr. Murray's knowledge?"

"Oh, I cross this field every once in a while. I never get into trouble; Mr. Murray's house is out of sight, over yonder behind the hill."

Donald indicated the direction of the residence with a wave of his hand. We crossed the last few yards of the field and stepped onto a broad bed of red sandstone bordering the quarry. Along the exposed bedding plane was a trace of worn ripple marks. The morning sun ricocheted red fire across the dark stone, highlighting a dinosaur track just as I was about to put my foot down on it. It was partly covered with a thin layer of debris, but under the leaves and dust there was no mistaking the outline. Donald watched as I brushed away the trash with a whisk broom. It was the track of a medium-sized carnivore, exactly like the first one I had ever seen in Arizona.

"There's bigger ones up in the main part of the quarry," Donald said.

I looked for the spot where the creature should have put his foot down for his next step. The track was about three feet distant, and the trail crossed the flat ledge to its disappearance in a weathered section.

Donald led me to a place where the rock had been more recently removed. Our shoes crunched sharply on bits of curling matrix, loosened by the exfoliating action of heat and cold. We entered a sloping pit which ran for some thirty feet back into the hillside. Here another trail crossed the flat surface underfoot.

"Well," I said, "it's hard to believe. Last time I followed one of these boys, out in Arizona, I never dreamed a couple of years later I'd be doing the same thing in Massachusetts."

"I think I remember more trails up higher in the quarry, just to the east," Donald offered.

"Well, this clinches it; we've got to look up Mr. Murray. I can't go back to New York without knowing more about the man who owns the quarry and knowing if he'd consider letting the museum have a few samples."

Donald grinned broadly. "In that case, we'd best start off right; we'll get in the car and drive in the front way."

Down the road a short distance a lane turned to the right. Ahead, a small white house nestled under a group of maple trees. As the car approached, a big middle-aged man appeared at the side door.

Typical Connecticut Valley dinosaur tracks, at Dinosaur State Park, Rocky Hill, Connecticut. Courtesy John Ostrom, Peabody Museum of Natural History, Yale University.

"That's him," Donald whispered.

We pulled up to the gate, and our man came toward us, with an inquiring eye. I got out of the car to meet him.

"Mr. Murray, my name is Bird," I said. "This is Mr. Simpson, who has kindly driven me out from town. I'm from the American Museum in New York and interested in dinosaur tracks. I understand you have some nice ones in your quarry."

Mr. Murray failed to note my extended hand.

"H'mph!" He glowered for a long moment. "Who do you know at the American Museum?"

"Well, I'm pretty well acquainted with Barnum Brown; fact is, he's my boss."

"Barnum Brown!" The name was magic. Murray suddenly noticed my extended hand and put it to use.

"So you work for Barnum Brown. Well, now, that's different. You must forgive me. For a moment I thought you were another one of those fool students. You must realize the spot I'm in; if I let every Tom, Dick and Harry in to see those tracks, I'd not have a bit of peace day or night. You can imagine!"

I tried. I found I could imagine. Well, I cheated a little; I remembered the visitors at Howe Quarry, and I told Murray sincerely he had my sympathy. I learned he had once met Brown at the museum. He immediately offered to take us on a tour of inspection, and we followed him afoot along the lane that led to the quarry. Donald and I examined the tracks as carefully as if we had never seen such things before. There was an abundance of trails, many of them in a fine state of preservation, some of them trails Donald hadn't seen before.

"There aren't any better tracks in Massachusetts," Murray boasted. "Look like you might be interested in a few for the museum?"

We discussed the problems of removal and financial considerations, but I explained I'd have to get Brown's okay and that I'd like to see whatever other tracks might be found about. I took several pictures, and after an hour's pleasant conversation we returned to the house.

"Well, boys, I hope you'll come back. And remember, you're welcome any time," Murray told us as we waved good-bye.

It was just past noon when Donald and I got back to town, and his mother insisted I stay for lunch. The rest of the family was as interested in fossils as Donald, and we chatted in lively fashion over the morning's events. Suddenly I recalled Miss Cooley's story.

"Donald, do you know Carlton Nash, out at Moodys Corners?"

Donald and the family looked blank.

"Well, it seems this Nash boy knows where a dinosaur sat down and left his mark."

The family laughed, and a thought occurred to me. I hated to impose on good nature, but here was a boy with a car and an interest in the subject at hand.

"How would you like to go with me in your car and see what he has?"

Donald was willing and ready. Soon we were rolling north from South Hadley. We had obtained some directions, and as we crossed a slight rise and saw a tree-lined brook ahead, we knew we were there. A large white frame house fit Miss Cooley's description of the Nash farm.

Beyond the bridge, what had once been a plowed field reminded me of a story in Lull's book. It was at Moodys Corners that Pliny Moody as a boy of twelve had plowed up a slab from the bedrock and found on it the first fossil footprint recorded in America. In 1812, a lifetime before dinosaurs were discovered, it could only be explained away as the footprint of Noah's raven.

Carlton's mother answered our knock at the door and was delighted when she learned I was from the American Museum.

"Carlton will be thrilled," she assured us. "He isn't here, but he'll be home soon. He just dragged his prize rock up to the house with his car. Would you like to see it?"

She led us down the walk to a large slab lying in the grass. It was a dark red sandstone, three or four inches thick, some six feet long. The surface had a varied array of imprints. Some were of small carnivorous dinosaurs, but in the midst was a larger impression. Leading away from it was a series of marks suggesting the print of a heavy tail.

"Carlton's worked his fingers to the bone cutting that slab from the creek," Mrs. Nash told us. "He'll be tickled to have someone from a museum see it and pass on it."

I could see how a neophyte might have thought that one of these little dinosaurs, only five or six feet tall, might have sat down. But as the specimen lay here, removed from its surroundings, much was missing to back up his theory. Where had the dinosaur come from? Where had he gone, when he grew tired of sitting? None of the tracks on the slab represented a trail that seemed to lead to or from the impression. If the print were really that of a sitting dinosaur, the bulk of the story still lay in the

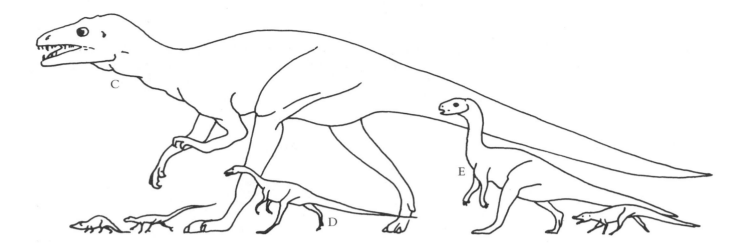

An interpretation of the kinds of animals responsible for the fossil footprints of the Connecticut Valley. The dinosaur tracks were made by the animals labeled C, D and E. Drawings by P. E. Olsen.

bed of the stream where the slab had come from. Or perhaps the mark represented only the spot where some object like a floating tree stump had rested on its way downstream while the mud was soft.

When Carlton showed up, I at once liked the serious young man who shook my hand. He explained he had found the odd impression in the bed of a small stream on a neighbor's farm and had gotten permission to remove it. I wanted to see the spot, but it was getting late in the afternoon, and Donald announced it was time he got home.

"You'll stay with us, won't you, Mr. Bird?" Mrs. Nash offered. "Tomorrow Carlton will take you over to Mr. Aldrich's."

At dinner that evening I met the rest of Carlton's family. Lenora, one of Carlton's sisters briefly home from a medical school in Michigan, was as enchanted with fossils as he was. Under her rapt attention, added to Carlton's, I blossomed with stories of my short and happy life as a dinosaur hunter. It was a rare thing to find, outside of museum people, other folks as interested in the world of science as I had always been.

Next morning, planning the trip, Carlton told us, "Mr. Aldrich doesn't much like people coming around to look for dinosaur tracks. He's bothered so often, he's to where he's about ready to shoot anybody who mentions the subject."

He told us this as we were coming down the walk from Mr. Aldrich's house, where we had found nobody home. The walk, a red sandstone walk, was interesting. Here and there an occasional flagstone was smudged with prints of three-toed dinosaurs. Carlton skipped on ahead, led by sounds

from the barn. Seconds later, he appeared from the barn accompanied by a big man in shirtsleeves whom he introduced as Mr. Aldrich. Seeing he was unarmed, I extended a hand and started to introduce myself, but Carlton had already filled in his neighbor on me. He shook hands with warmth, and we talked about the tracks along the walk. Aldrich led us to a sizable pond above the house, where a creek had been impounded to run an ice plant a bit down the valley.

"When we were digging the ditch for the flume," he said, "we cut into a ledge covered with tracks."

Carlton brought up the principal object of our visit and asked if he might show me the place where he had cut out the slab.

"Sure, Carlton; take Mr. Bird anywhere. The place is yours," Aldrich replied. Apparently the barks of dinosaur track owners in these parts were, on the evidence to date, worse than their bites.

In the creek bed the evidence we found only served to blur further the sitting dinosaur picture and to reaffirm my opinion that some inanimate object had made the strange imprint Carlton had quarried. Much as I would like to have seen things the way Carlton hoped, I couldn't quite go along with his interpretation.

Not that his idea was unreasonable. Weeks later I found in the museum at Amherst College a set of prints collected years before which clearly told a story of a bit of a day in the life of a dinosaur. He was a little fellow, walking upright on his splayed little feet on a mud flat. His hind feet made perfect prints in a pliable matrix. A shower came up; large drops of rain began to fall, making little dents in the mud like falling shot. They pelted the little fellow and he hunkered down, touching the toes of his little forefeet to the ground. His posterior sank into the soft mud, as well as a bit of his tail.

The shower passed; the sun came out; the dinosaur went on about whatever had been on his little reptilian mind before the shower intruded on his thinking. The record is clear there in the Amherst College museum basement; it would stand up in court. Raindrop impressions, fossil spatter-marks, show clearly in his tracks up to the point where he had stopped to sit out the shower. His body print and the tracks that continued across the pelted flat stand clean and clear.

Skeleton of *Monoclonius*. Courtesy Department of Library Services, American Museum of Natural History.

14

A WARM SEPTEMBER afternoon. I was at work refining the job on some incompletely cleaned bones. Otto Falkenbach stopped by my desk.

"Charlie Lang says he's found something downstairs he wants looked at. Want to come along?"

It was a welcome suggestion. Museum work was never dull, but when a job goes on long enough, it gets tedious.

Otto picked up a can of shellac. "Better bring your brush and cup," he added.

We took the elevator to the basement. Otto led the way along a lengthy cavernous passage under the welter of pipes that supply the museum's life blood—water, steam, electricity—to the floors of this tremendous red barn. Charlie was nowhere in sight. We came to a locked door. Otto's keys took us into a great dark room smelling of musty straw and long-undisturbed dust. He found a light switch, and the bulbs revealed piles of unopened boxes stacked about the walls. Under thick coats of dust were labels, names and dates of expeditions many of which had taken place before I was born. Boxes close by bore the label, "Belly River Cretaceous," a name I associated with Brown's work in Alberta long ago. Others were tagged "Bone Cabin Quarry," where Brown had unearthed the first brontosaur in 1897. Otto looked around carefully.

"Not here," he said. "Charlie must have taken it to the room across from the old print shop."

I followed him back through another of the mile-long passages that were part of this underground labyrinth. We passed before another door, and again Otto's keys let us through. The room we entered was partly lighted by high windows, but back in the shadowy corners it was difficult to see. Otto flicked on another switch. A framework of iron supports holding large removable racks ran from floor to ceiling. The racks were loaded with hundreds and hundreds of dinosaur bones, most of them duplicates of skeletons on exhibition in Dinosaur Hall. Many were simply too incomplete to merit mounting. Here was stored an extra skeleton of a tyrannosaur, the great limb bones and the teeth of the lower jaw just visible over the edge of a rack. Here were the dorsal vertebrae, as big as washtubs, from a brontosaur. The skeleton of a mosasaur, a great sea monster, lay dismembered nearby. All in all, there seemed to be more material here than in all the upstairs. Otto still didn't find what he was looking for.

"They must have taken it to the new Dinosaur Hall after all," he said.

We set out again on our underground walk, presently coming to the basement of the new Roosevelt Memorial Wing. We took the elevator to the fourth floor. The new Jurassic Dinosaur Hall stretched before us. Brown planned soon to move all the Triassic and Jurassic dinosaurs into these new quarters, leaving those of the Cretaceous . . . dinosaurian twilight time . . . in the old and now quite crowded hall. But at the time, the room was being used for storage. Down at the far end, loaded on platform trucks, we saw three or four large wooden boxes from which the tops had been removed.

"This must be it," Otto said.

We crossed the tiled floor, our shoes clicking loudly in the silent room. The boxes contained large blocks of sandstone, from which plaster and bandages had been partly stripped away, revealing bone buried in the matrix below. A whole animal, but for the skull, seemed to be here, the bones perfectly articulated. A label lay in each box. The label read, "*Monoclonius* skeleton, collected Alberta 1914." The bones were in fine condition but dried out from long standing. Otto opened his can of shellac, and we set to work flooding the thin fluid along the bones.

"This is a very fine thing everyone had forgotten was here," Otto remarked.

For the first time in years, Brown had begun an inventory of old material. The museum planned to dispose of some of its surplus by giving it to schools or colleges. Skeletons of excellent exhibition value, such as this *Monoclonius*, could be traded or sold to other institutions. We had a fine *Monoclonius* already on display in Dinosaur Hall. The creature was a close relative and quite similar to *Triceratops*.

A day or two later Charlie and Jerry came into the lab wheeling a truck loaded with four or five trays of wrapped packages, which they piled on Otto's table. Some of the packages were open, their contents revealed. I read one of the labels still stuck to the wrapping; these were the fragments of skull of the specimen Otto and I had shellacked the few days before.

"You and Otto can try your luck on fitting these back together," Charlie said, "sometime when you're not busy. Brown wants to know if there's enough of the skull here to be worth salvaging."

Otto and I began unwrapping the rest of the packages. The fragments were hard and firm with clean-cut fractures.

"A lot of this will go together easily," Otto noted. "Let's spread 'em out here and see what we have."

According to the label, we could expect another *Monoclonius* skull, three to four feet long with a great, parrot-like beak, an impressive horn projecting from the center of this beak, just in front of the eye sockets, a great standing frill of bone forming the back of the skull and extending back over the neck and shoulders. We picked a number of the larger masses that were obviously from this upstanding frill and worked them into place. But in several places where there should have been a smooth edge to this frill there seemed to be broad bases for sizable horns.

"This can't be *Monoclonius*," Otto said. "*Monoclonius* never had horns up here."

We went on with our stone puzzle, coming upon a few sections of horn that slipped into place on the edge of the great frill. When Brown came in to see what and how we were doing, his nonchalant viewing turned to sudden amazement.

"This is no *Monoclonius*," he exclaimed. "We've got a *Styracosaurus* here. See that crown of spikes . . . or you will see, if the rest of the spikes are here."

He began rummaging through the remaining stone bric-a-brac. "This is a find. The only whole *Styracosaurus* skull in the world is in the Canadian National Museum in Ottawa."

He hurried away, returning with the literature on the Ottawa specimen. For the next hour and a half the three of us worked together, assembling more of the frill and beginning the underside of the skull.

For the next two days we went on finding contacts with ease, Brown stopping by from time to time to lend a hand. By the third day, work began to slow down. We had added to the frill and had put together enough of one side of the skull to place the quadrate, that element of a reptilian skull from which the jaw is hung. But by now we were running out of easy-to-find contacts. We now began to count each contact we made, putting tally marks on the paper that covered the table. When four o'clock came, Otto had marked up eighteen for the day. I trailed with fifteen, but this was my first big bone-piecing job, and I felt good. The work was simply a great jigsaw puzzle. The fact that it was three-dimensional and with many of the pieces missing made it more puzzling. Otto and I worked on that rare skull until the job became stale and results fell to zero. When we came to a day when neither of us found a single contact out of the hundreds of pieces remaining, we decided it was time to turn to something more productive.

"At that, this isn't a bad job compared to what Jerry and Carl did on that first armored dino from Montana. They worked on it two years straight."

A commentary on the ways of science and the nature of great museums as well as of man in general: in the course of Brown's inventory, cases were moved from behind a screen on the fourth floor. Leaned up against the wall was a great slab of red sandstone, covered with dinosaur tracks. The label told us it had come from Turners Falls years before.

"This solves the problem of a display of tracks from New England Triassic," Brown pointed out. "These are better than any we saw last summer."

Otto went off to have a try at planning a mount for our skeleton, and I went back to the job I had left for the skull. A few days later we chanced to meet again at the job we had given up. Still there were trays full of odds and ends and bits and pieces we couldn't possibly find a place for. Idly Otto tried a few pieces together. They fit! Refreshed by a few days rest, in a few minutes he came up with five fits from the trays of trash on which we had given up. Cemented together, two of these fit into an empty place in the skull.

We had difficulty making an accurate estimate on the overall length of the skull; too much was missing from the snout. Carl and Jerry tried a hand now and then as they went by. Brown occasionally fiddled a bit with the leftovers. A young instructor at Brooklyn College, Dr. Erich Schlaikjer, stopped by from time to time, doing no better than the rest of us. Then one day he picked up a long sliver of bone at the edge of a tray, a random piece nobody had touched for a good while. He placed the tip at random against one of the already established jaw elements. It fit beyond question.

"What's the matter with you fellows?" Erich inquired. "Here's practically the rest of your jaw plan right before your eyes."

We happily conceded he was right. Moreover, this established contact reminded me that the fragmentary end of the parrot-like beak lay in another tray. I tried it for contact on the other end of Erich's sliver. It fit the tip of the element, called the predentary, establishing at last the length of the skull with exactitude.

A few days later Otto cleaned the jaw that had been pieced together. Though still incomplete, all eight elements of the ceratopsian jaw—the jaw type of the *Triceratops* and their kin—had been accounted for. One suture, however, seemingly extra, appeared at an unexpected point. Where only a single element of a proper reptilian jaw was supposed to exist, there were two. Otto cleaned the joining edges with his accustomed thoroughness and care. After considerable work with a fine needle under a microscope, he separated a small flat bone from the rest of the mass.

"I don't understand this," Otto said excitedly. "This must be an element no one has recognized before. What do you make of it?"

Just as surprised, I got out a book showing all the known elements of primitive reptilian jaws. Charlie Lang came over to the table, looked, wondered. Brown came by, was given the news, pored over the evidence. The identity of the newly discovered jaw element was no longer in doubt; what remained was learning if this was standard in the creation and formation of reptilian jaws.

To the man in the street, this matters not in the least. Most people recognize only two parts of the head: jawbone and skull. Students are often surprised to learn it takes a couple dozen bones to make a working skull, and even the jawbone is not a one-piece job. Mother Nature worked a long while shuffling all the pieces to make the head of all her creatures, and among the reptiles most of the parts are poorly fused. Reptiles, in fact, got such a poor job on jawbones that the snake, best-known of the whole bunch, has such a poorly constructed jaw he generally throws it out of joint when he wants to eat. The jawbone of mammals was much better constructed, and as mammals mature, their bones generally become more fused together. The reptilian jaw is a different case.

Coming upon a separate element of a reptilian jaw unrecognized all these years was a remarkable find. We checked every ceratopsian jaw in the museum. We borrowed jaws from Carnegie, from Peabody at Yale, from the National Museum in Washington. In every specimen the intercoronoid element was brought to light.

Come spring, talk of another expedition sprang up. Brown was jubilant; after two long years, it looked as if we might get out West again. "We've been assured of some money from the National Park Service," he said. "Padded out a bit with aid from here and there, we should be able to get away by the end of June."

We began to lay plans as though the trip was already assured. This was to be no large party, like the Howe Quarry thing; Brown and I would first visit the Dinosaur National Monument near Vernal, Utah. He was to study the situation and to make recommendations to the Park Service concerning future development. Afterward we hoped to go on to other prospects and, last but not least, we hoped to revisit the area around Cameron, Arizona. Mrs. Brown had been on a lecture tour during the winter, using the company Buick, but now the old green touring car stood in our barn at Rye, ready for use.

I had heard a great deal about Dinosaur National Monument. Within the park's borders was one of the richest dinosaur areas in the United States. A great number of creatures had been found there. Carnegie Institute and the National Museum, as well as other institutions, had gathered material here from time to time. Both sides of the great deposit had been worked, but the intervening portion, now exposed by a great cut in the grey Morrison sandstone, was presumed to bear more skeletons. So far, there were no bones of importance in sight.

"All the skeletons are articulated," Brown explained, "and lie under an overburden of several feet. When they are exposed, they will be left in position on the face of the sandstone. The final exhibit promises to be unique."

In the massiveness and abundance of its fossil material, the display would be similar to the Howe Quarry, but as the skeletons would be exposed in sandstone instead of clay, they would stand exposure much better.

Unlike Howe Quarry, here the structure was sharply tilted, so the display would be observable to viewers as a great wall with dinoskeletons plastered on it, ideal for good viewing. The two-hundred-foot wall would bear in deep relief the great sauropods, the bipedal dinosaurs, even some of the little fellows of the Jurassic Era.

But we are jumping the gun; back to the museum.

Late in May we had cleared Jurassic Hall of the material stored there during the winter's inventory of mislaid, undisplayed, forgotten bones. Now talk of moving in the new occupants began. I was still helping Otto with the *Styracosaurus* when Charlie Lang got out great coils of rope and block and tackle, and for a few days he was missing from the lab but for such times as he dropped in for more rope or other paraphernalia.

On the third day Otto announced, "I hear Charlie's going to move the *Triceratops*. Let's go see how he does it."

Triceratops was still frozen about to lunge at an adversary, there in the old Dinosaur Hall. But now he was the center of a great bustle. With men working under him, around him, by him, he had never loomed so huge. The great skull with its enormous frill, a combination of neck piece and shoulder guard, seemed big as a bathtub. The great framework back of it with its four widely spraddled legs loomed like a lumbering boxcar. The base of the mount was wood, and the framework under it rested on the tile floor on castors. Charlie, Carl and Jerry, fitting block and tackle to the front end of the mount, saw us coming in the door.

"Just in time," Charlie called out. "Come and give us a pull."

He handed us the end of a rope which passed through the block and tackle lying on the floor. *Triceratops* was not to leave the room; he was simply to move forward in the hall. The others took up the rope, and Charlie gave the order to heave-ho.

The rope tightened. The wheels of the blocks creaked. For the moment the popping ropes pulled at a dead weight, a weight that dragged its feet, refused to budge. Then gradually and grudgingly the base of the mount began to move, a groaning and rumbling juggernaut. *Triceratops* in motion was a different picture than I would have imagined; he was not pushing forward in a ferocious charge, but shaking in every joint and bone. The great skull rocked ridiculously from side to side, on its pedestal of iron rods. "Old Tricky" didn't want to go. Charlie peered anxiously up among the quaking ribs for any sign they might be coming apart.

Slowly an expanse of floor, hidden more than a decade, peeped out from under the mount. It was covered with dust half an inch deep, littered with pencils, gum wrappers and a baby's teething ring.

"Late Cenozoic litter," Charlie commented. "Late . . . very late . . . Man. Well, no; the baby rattle and the teething ring would be very early man."

A week later Charlie paused at my table. "R.T.," he said, "we're ready to move 'Old Bronty.' You thought we needed a hand before? Well, this time, we need all hands."

Downstairs, the grand old *Brontosaurus* didn't look so grand. He was missing three legs and a tail. Above what remained of the skeleton, the workmen had rigged a trestle to carry block and tackle for lifting bones off their metal supports and lowering them to the floor. Jerry and Carl were tying ropes to the right femur.

"This last leg he's standing on will be easy," Charlie said. "But when we come to the pelvis . . ."

The femur, weighing five hundred pounds, was lowered from its position off the right hind leg and brought to rest on top of a waiting platform hand-truck.

"How did you ever handle the tail and its base?" I asked.

"No sweat getting it into the freight elevator here. No trouble getting it across the courtyard downstairs. But we hadn't an inch to come and go on in the freight elevator in the other wing."

I looked to Charlie for instructions; after all, he hadn't run me down just for conversation.

"We got to get some rope, plenty rope, around the pelvis," he said. "You want to see how things look from the top?"

Jerry steadied a twenty-foot ladder, while I

climbed up carrying a rope. Charlie went up a ladder on the other side, held by Carl. From my perch the floor looked a long way off, but I kept my attention on passing the rope under the brontosaur's hips, an eighteen-hundred-pound mass of petrified bone, to Charlie's outstretched hand on the other side. Then Charlie swung the rope over the top, over the neural spines, sending it back to me. In reaching for the rope I partly lost my balance but caught hold of one of the neural spines to steady myself.

"Careful of that pelvis; don't lean on that spine!" Otto called from below. "Remember: we can always get another man, but that's Barnum's

first brontosaur." Brown had dug this fellow up at Bone Cabin Quarry in 1897.

When we got the pelvis down, it looked fully as large on the truck as it had looked right under my nose from twenty feet in the air.

"Now," Charlie said, "when we get the upper tailbones and the backbone down, we can bring in the other sections of the base and we're home free . . . all but putting the old fellow back together again." Putting the old fellow back together in his new position in the same old quarter took sixteen days.

Our western trip, postponed time after time, was finally set, the funds secured, and a day picked

Skeleton of *Styracosaurus*. Photo by Charles H. Coles and Thane Bierwert. Courtesy Department of Library Services, American Museum of Natural History.

to go. But just before we left, a box came in from Texas. It contained a large fossil palm leaf, not quite complete at the front end, but otherwise in fine condition. It had been collected in the Big Bend country in southwest Texas by Don Guadagni. I had met Guadagni a month or so before, an amiable fellow who had dropped into the museum for advice on fossil-hunting on his vacation. He had done a little collecting from the Paleozoic rocks of New York, but aside from this experience with invertebrates, he was entirely new at the game. Brown had suggested the Cretaceous of west Texas and had given him a lead on this fossil leaf, reported in a letter from a rancher in the area. Brown was well pleased with the big leaf.

"Guadagni plans to join us for a few days when we get to Cameron," he told me. "He should make an excellent collector."

We left New York the first of August. Brown had to call on the Park Service people in Washington before heading west. Then we detoured slightly to visit the old home of President Monroe, now owned by a Mr. Littleton.

"Littleton has opened a quarry in which there are some fine dinosaur trails," Barnum explained. "He has promised some of them to the museum if a suitable slab can be located."

Oak Hill, as the estate is known, is in Loudon County, one of the few locations in Virginia where Triassic is exposed. As we rounded a corner at a country crossroads, the old colonial mansion stood out clearly on an oak-crested ridge. Shortly we turned off the highway following a long lane lined with huge oaks. The lane led to an entrance on the north side of the house. Lawns stretched away in all directions from a circular driveway of dark red, fine-grained, thin-bedded stones.

Several stones in a nearby wall contained ripple marks. The third stone I looked at showed rows of short scratches, finely defined, about a quarter-inch apart. It was the trail of a little insect, preserved in the surface of the rock. A Triassic bug or beetle had dropped on the soft mud here, something better than two hundred million years ago, and left us a message that he was headed somewhere, written finely in his tiny footprints.

When we found Mr. Littleton, he led us back to the house, through the north portico. The shades on the high windows were partly drawn, but in the dim light I could make out fine furnishings of another era. Our host explained that when Monroe was president he had built this place, a

day's ride from Washington by carriage, to serve as a summer home. Traditionally, it is supposed that the Monroe Doctrine was conceived here.

When Mr. Littleton led us out onto a broad verandah on the south side of the house, I was intrigued by the large flagstones, red like the stones out in the wall. Our host turned to a door on the right and pointed down.

"It's amazing how many years it was before anyone recognized this as a dinosaur footprint. After all, dinosaurs weren't discovered until Monroe was dead and gone."

I was taken by surprise. The imprint at our feet was familiar; it might have been in Arizona instead of here . . . or in the backyard of Aldrich's place in Massachusetts . . . or Murray's Quarry . . . or outside Cameron, Arizona. The shape of the three toes and the scratch-marks of the claws were easily recognized, but someone had obviously not been pleased by their appearance.

"One of Monroe's slaves apparently did his best . . . a poor best . . . to scrub out the prints, considering them a flaw in the flagstone. A little deeper, and he might have scrubbed it all away."

I looked at Brown, for his reaction. He smiled and nodded; he had known about the print all along. Littleton had bought the property only a few years before, and it was he who had first recognized the print for what it was.

"The biggest mystery of all was where this stone had come from. After I bought this place, after looking all over Loudon County trying to trace it down, I found the quarry it came from was on the property," Littleton told us.

Three days later we were crossing the eastern part of Barnum's home state, Kansas. Here as a boy he had found his first fossils in the ancient Paleozoic limestones cropping out by the wayside. Rolling into western Kansas, we watched for a first glimpse of the Rockies; it seemed ages to us both since we had seen them. They first appeared as little blue clouds on the far edge of the flat plain ahead. Then they were no longer clouds, but patches of a deeper blue, too land-bound to be clouds.

We spent the first few days running down minor leads on our itinerary. We chased a dinosaur rumor at Durkin Ranch . . . only a rumor, no bones.

Two days later, across the mountains, we entered Vernal, Utah. The surroundings of the Dinosaur National Monument were spectacular, folds of Permian and Mesozoic rock cut through by the

Green River at Split Mountain. Work on the Park project was coming along slowly, with no bones to speak of as yet in sight. Brown spent days studying the progress that had been made and preparing his report for the Park Service.

We spent a day or two prospecting the Morrison Formation, greatest of all dinosaur repositories in the United States, east of Vernal. Then we headed south in Utah. The geology, to me, was no longer simply pictures without labels, as it had been when I had first come here all alone; now I had a fair knowledge of what I was looking at. We visited Zion Park and Bryce Canyon. East of Bryce we dropped down on the hot desert in search of another reported dinosaur. This time the bones were real, but they were in an area that couldn't be approached by car. Byron David, the boy who had written Brown about the dinosaur, wanted us to visit the spot on horseback, but our time was too limited.

Next evening we rolled into Cameron, Arizona. The little Indian trading post huddled close to the big steel bridge looked familiar even in the dark. Hubert Richardson, the trader, and his wife had been looking forward to Brown's arrival for several days. We signed the register in the lobby and were invited to join the family at dinner. Richardson didn't remember the boy with the motorcycle who had once asked him the way to Dinosaur Canyon, but during dinner he and Brown reminisced over

old times. Plans were worked up for the following day. Brown decided we would visit the locality from which *Protosuchus*, the ancestral crocodile, had come. "I'm going to take the day off and go along with you," Richardson said.

After the meal, on our return to the lobby, a young woman arose from the shadows of the lounge and approached Brown with a smile.

"Pardon me, but isn't this Dr. Barnum Brown, of the American Museum?" she asked.

Brown beamed all over, admitting her guess was correct. The lady said she had seen his name on the register and felt sure we were here on one of those exciting expeditions of which she had read. "Could you possibly tolerate our presence . . . mine and my husband's . . . tomorrow, while we watch you work?"

"We'll be delighted to have both of you," Brown chuckled. He made it sound as if our day would be duller, our chances of turning up anything poorer, if they didn't come along. The lady was Adelle Davis, a consulting nutritionist from Hollywood, here on a vacation from prescribing diets.

As we started to leave the coffee shop a very large man, with smile to match, just walked in the front door. Don Guadagni.

"Looks like the gang's all here," he said, offering a big hand.

Skeleton of *Triceratops*. Photo by E. M. Fulda. Courtesy Department of Library Services, American Museum of Natural History.

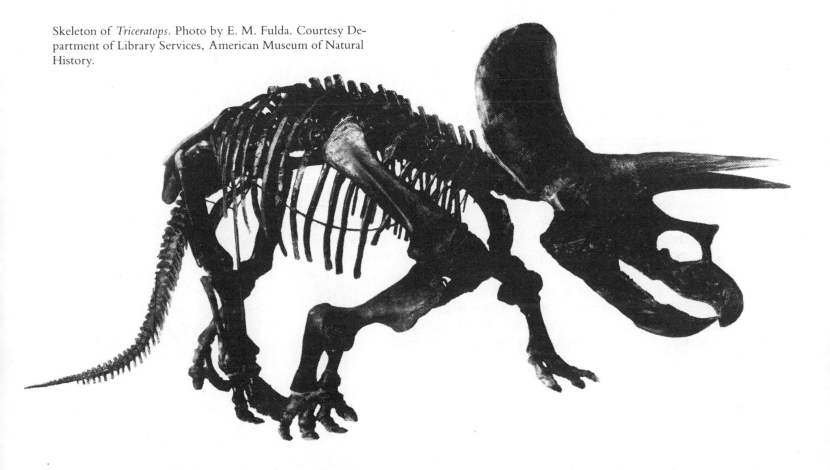

15

THE SAME BLAZE of dawn that broke over the Painted Desert east of Cameron splashed over the wall of my room and brought me out of bed. I went to the window and looked out on the canyon of the Little Colorado. The wall opposite dropped from the rim sheer and straight, like the side of a red box. The distant horizon across the desert was still a dark line of cliffs against the rising sun. I dressed in a hurry and knocked on Brown's door.

In minutes we joined Don Guadagni downstairs and went in for breakfast with the Richardsons. In the lobby we found our Hollywood friends ready to follow us to the field in their car, but Brown suggested they ride with one of us.

With gassing up, checking tires, and unloading Don's car a bit to give him room to bounce around on a rough road, it seemed we'd never get off. With it all, we still got on the road by half past eight. Hubert led the way in his big Packard, with Brown as passenger. I followed with the company Buick, carrying Adelle Davis and her husband. Don in his little Ford came on all by himself.

We negotiated the first bench and a series of ridges beyond and dropped into a big dry wash. Sand flew back from the Packard's wheels. The Buick's wheels churned the soft sand like paddle wheels. In the mirror I could see Don's little Ford chugging along behind. Don was grinning broadly; he had experienced this sort of driving in the Big Bend country and realized its trickiness.

The wash swung around a bend, widened, narrowed, began to fork in the face of a rising bench. The going was neither better nor worse than on the day of my first wild ride here, but a good bit easier to handle than it had been with my old Harley and its cumbersome sidecar. Surviving all hazards of rock and sand, the procession finally climbed out on one of the low benches below the cliffs near Dinosaur Canyon. The weathered rocks with their leering faces stood all around us as Hubert and Brown bumped to a stop. So stopped we all.

Brown climbed out of Hubert's car, picked up his tools, and headed for a low hill of red sandy clay, followed by the rest of us. There were no signs of digging, but here, Brown said, was the place from which the original *Protosuchus* had come. He explained that we would work around the sides of the hill, in hope of finding another of the little eighteen-inch crocodiles. Our guests were given crooked awls and a paint brush each but cautioned to go slowly if they ran into any sign of fossils. "The bones will be quite small, chalky white, and soft," Brown told them.

We all worked quietly for an hour or so. The sun got higher . . . and hotter. But once, when I rested my eyes by turning them away from the glare, I noticed a tiny plume of cloud beginning to form around the highest peak of the San Francisco Mountains, sixty miles to the southwest.

Brown and Hubert uncovered a few tiny bone fragments, *Protosuchus* ribs. The small discovery stirred everyone to fresh efforts. Brown carefully cleared an area around his find, but there was no more of the little fellow's skeleton to be seen. Presently he sighed, picked up his tools, and suggested we stop all this for a bit and prospect instead for sandwiches.

"There's no use spending more time on this hill," he said. "Maybe we can find a better spot after we're fed."

We got our lunch boxes from the cars and withdrew to the shade of one of the nearby rock formations, where a hideous grin etched across the face of an opposing pillar mocked our morning's efforts. As we found seats and leaned back against the rocks and started to eat, a muffled sound reverberated across the desert, like the rumble of a distant cannon. We looked toward the sound, toward the San Francisco peaks.

A formidable thunderhead filled the western sky. It had built up around the mountains until they were completely hidden by the ominous black cloud. As we watched, it drifted eastward across the desert. We stared, fascinated, as a sheet of rain drew a curtain across the benches beyond the Little Colorado.

"Let's scram outa here," Hubert said, uneasily. "She's headin' here sure as shootin'. An hour from now there'll be water three feet deep right where we're settin', even if we're standin'."

We wolfed the rest of our sandwiches as the advancing thunderhead blotted out the sun. A flash of lightning split the rain curtain and thunder came on close behind it like the crack of a whip. We piled into the cars and began to turn around. In the haste to get away Brown piled in with me, but Hubert led the way as before. The course of the sandy, winding dry wash led directly into the heart of the storm. Behind Hubert, ahead of Don, the Buick lurched through a stretch of sand with laboring motor. Brown pushed forward against the dash with both hands, as if to push the car forward. "Pour it to her, R.T.," he urged. "Pour it all to her!"

The Buick wallowed and galloped through the loose sand like a speedboat in rough water. We rounded a long bend, approaching the spot where the desert trail had dropped us this morning into the wash from the north. Hubert swung his Packard up this path, followed closely by the rest of us. Just as Don's Ford chugged over the rim the rain came down hard.

We paused on the rim to wait for the rain to pass. After an interminable time it slackened but didn't stop.

"This is the place where I found Mother Carey's Kitchen back in the days when I was still a hairy mammal. I'm going to do a little prospecting," Brown said, after a few minutes, and started off in the light shower. Mother Carey's Kitchen was the name he had given to a find of old clay pots and various other Indian artifacts he had dug up here years before. I pulled on a coat, pulled loose another shovel, and set out after him.

"What's with Brown?" Don inquired. "Now if we were going duck-hunting . . ."

Even our Hollywood guests decided to string along, reluctantly, I felt. After all, Brown didn't have to worry about getting his hair wet.

About a hundred yards away, Brown started digging, coming upon some flat stones stood on edge. "The meal bin was right here," he told us. We all set to work digging, turning up charcoal and ashes of long-dead fires, bits of pottery. Nothing exciting but interesting enough for us not to notice when the rain stopped. We did notice, however, that the wash we had come out of was no longer a dry wash . . . it was a raging flood. It would be hours before we could cross it and return to Cameron.

"If we drive down toward Tanners Crossing," Barnum told Hubert, "we might have another look for fossils. Looks like a dry hole here, rain and all."

We followed Hubert into a grey area that seemed stratigraphically in the banded lower Triassic. Rounded hillocks, thirty to forty feet high, splashed with maroons and purples, appeared on the left. All three cars tiptoed apprehensively through some very liquid mud, but nobody got stuck. We closed in on the area Barnum wanted to prospect.

"I took out some phytosaur limb bones here in 1903," Barnum told us. "In the meantime, there's been some erosion. Should be worth a fresh look."

We all joined in with new enthusiasm. The rest of the party found bone scraps lying about here and there on the bare and eroded surface. Before he had covered a dozen steps, Brown picked up a fossil tooth, long and conical, like a crocodile's. Moments later he picked up an odd lumpy object. The tooth was easy to associate with a particular type of creature, but the lump puzzled the rest of us. It was elongated, stony of course, pinched off at one end. We passed it from hand to hand.

"That is a coprolite," Brown explained. "Fossil dung." He presented the piece to Adelle. "You're in the nutrition business; science has learned all it knows about the food habits and nutrition of ancient creatures largely through microscopic study of these. They are fairly common in some horizons."

We scattered out and began to look over the ground individually. I picked up a bone fragment I was able to recognize as part of the skull of a labyrinthodont, a type of Triassic mud-puppy. Don and Hubert were finding a variety of bone fragments.

Our guests were thrilled to pick up sign. Nowhere had I seen a spot so sprinkled with Triassic odds and ends.

I swung southwest a bit, to avoid territory being gone over by Don, and followed assorted bone fragments that had weathered out back to their point of origin, where I began serious digging. I found bone buried shallowly in matrix and set to work on it with care. Looking up after a bit to rest myself, I glanced about to see how the party was doing. Don was immersed in a job of his own. Barnum, with Hubert watching, was working seriously on an open flat a few hundred yards away. Our Hollywood friends who had been idly surface-prospecting a few hundred yards off drifted toward Barnum and Hubert. I couldn't help but wonder what Barnum had come upon.

Adelle saw me looking and waved me to come over, join the group. I turned back to my specimen, thinking I had best finish it first. Part of a battered limb bone, difficult to determine what it belonged to. I saw Don desert his find in disgust and wander over toward Barnum and Hubert. Hubert was now shovelling too. Don waved me over. I joined the group.

Brown and the other three men were bent over a sizable excavation. In the bottom, over their shoulders, I could get a glimpse of the side of a great jaw, and a portion of a huge skull! The specimen looked to be five or six feet long.

"If this is a phytosaur skull," Brown said, "I've never seen a bigger one."

I swelled with pride in Barnum's success; he had never yet failed to bring back something big or important when he went out into the field, and here he was doing it again. The largest skull in the museum was not over three feet long. I picked out a favorable location at the back of the great jaw and went to work. The covering matrix was a hard blue clay that broke up into sharp, irregular fragments. The bone was hard but badly fractured. We stripped matrix, paused to apply shellac, stripped away again. The others worked to remove overburden.

Time flew. Shadows crept uninvited into the valley of the Little Colorado, fingered their way through the low blue clay hills. Little by little the skull was uncovered, but the overburden was heavy, and cleaning the bone went on slowly.

"I haven't figured this sucker out yet," Barnum told us. So far, the snout was still covered with hard clay. If it proved to have nostrils at the end of the snout, it would be a crocodile. But crocodiles were unknown at this horizon. If, on the other hand, the nostrils were farther back, just in front of the eyes, we had a phytosaur on our hands. As the sun dipped below the horizon, Brown began to pick up his tools. "We'll just have to let him rest here one day more. After all . . . after a few hundred million years . . . what's another day? Besides, we still have to find out if we can get home." We covered the hole with canvas, and started for the cars.

It was almost dark when we arrived at the big wash. Ahead, a stalled truck sat in mid-channel, its floor boards awash. Apparently it had been there a good while. It was one of Richardson's own trucks, from the post. The lone figure on the opposite bank was Bud, one of Richardson's boys. He waved.

"Where the heck have you guys been?" he called out.

Hubert snorted. "What are you doing in the middle of the wash?"

"Mom sent me. By now, supper's gettin' cold."

We watched Hubert attempt the crossing with the Packard. He got successfully around the stalled truck and boiled through the water to the farther shore. Don followed. The far bank was wet with water from Hubert's passing and splashing, but finally the little Ford chugged its way to the top. Bud climbed into the Packard and his father talked about the truck, deciding finally to come back for it another day.

Brown waved me on. I hurled the heavy Buick into the wash, lurching along in the ruts made by the previous two cars. We passed the partly sunken truck. The headlights skittered and wavered across the water, tilted skyward as we hit the bank. Then we oozed slowly back into the stream.

I was panic-stricken. I pushed the throttle well into the floorboards; the motor roared lustily; the Buick continued to slip downstream. I had failed to gather enough momentum to make the bank.

Then the tires caught on an upturned stony ledge, bit in, and we leaped to the top of the grade, the engine spitting and choking.

An hour later we all dragged into the trading post, looking more like ditch-diggers than science students and their devotees. Mrs. Richardson met us with a grim look, dropping a sarcastic remark about the kind of people who didn't know enough to get in out of the rain. "And how do you think I felt, when I sent Bud out to look for you and he didn't come back?"

Barnum laughed. "Why, Maud," he consoled her, "haven't I always told you never to look for a fossil-hunter until you can see him?"

16

THE HUGE SKULL turned out to be a phytosaur, the largest ever found. Teeth like marlin spikes. A jaw fifty-four inches from hinge to tip. We uncovered a few vertebrae, too, but nothing more; apparently the skeleton had been disarticulated before burial. Brown decided the limb bones he had found here years before must have belonged to the same animal; they were to the same scale.

It took two days and a half just to get the big skull ready for plastering and another day to get it properly jacketed. The job was particularly difficult because of the loose, dry matrix; this necessitated running preliminary bandages completely around the mass before we dared turn it. When we got it ready to haul back to the post, we were afraid to carry it over the rough trail in the Buick's trunk rack, but we managed to get it into Don's Ford by removing a door. Boxing it required another day, which ran us to the end of Don Guadagni's vacation.

For some days Brown and I continued prospecting the Arizona Triassic. We visited a spot north of Cameron where years before Brown had found the stumps of fifty or so fossil trees standing in the position in which they had grown, with fossil roots buried beneath them. This is practically unparalleled; in other localities fossil trees are almost always found recumbent.

Over in Winslow we stopped to look at the butte where I had stumbled upon the skull that had opened the museum door to me. We circled the butte, looking for signs of newly weathered specimens, but found nothing. "Lightning seldom strikes twice in the same place," said Barnum.

We chased about the Petrified Forest a bit, picking up a few impressive *Calamites* stems. We know this plant today best for its insignificant descendent, the horsetail rush, but once this thing grew to a height of thirty-five to forty feet and contributed materially to our coal deposits in some areas. We turned north into Colorado on a rumor that a road crew was scooping out fossil bones in a highway cut, but this came to nothing except to point out Barnum and I had a similar weakness: we knew the time was coming for us to head back for the barn, and we knew that the straightest line is usually the dullest distance between two points. The phytosaur skull was of course a notable find, but we wanted to add to it if we could. Heading due east next was somewhat against our principles, but Walter Granger, Curator of Fossil Mammals, had suggested we check out a lead on a Pleistocene mammoth reported by a farmer near Argos, Indiana. The farmer, a man named Thompson, had plowed up a tooth, struck another bone, and suggested by letter to Granger he thought there might be more bones there underground. Barnum wanted to take time out to visit relatives, and he suggested I drive the Buick on through and look the ground over, doing whatever the situation justified.

I arrived in the little Indiana farm town on October first. It was a late summer; the trees were only beginning to consider changing color; hay was of course all away, but most of the corn was still out, awaiting silo-filling. Mr. Thompson's farm was about a mile out of town, and I located him easily at the barn. He was an amiable man, quite willing to do all he could to help the museum recover the skeleton. He had been expecting me, he said, as he led us across the fence and into the corn field where the bones lay buried. The corn was a good bit taller than our heads. We pushed our way blindly through this leafy forest until we reached a row Thompson seemed to recognize.

"Right about here," he said, "I plowed the tooth out first. But there should be some other pieces of bone about, too."

Brown and Bird excavating the skull and partial skeleton of *Machaeroprosopus*, a large phytosaur, in northern Arizona. Courtesy Department of Library Services, American Museum of Natural History.

We turned to search along the row. Here was indeed a new experience in prospecting for prehistoric animals. The soil, instead of the hard matrix we usually worked in, was coal-black peat, decayed vegetation that had accumulated in a post-glacial swamp. This swamp had existed not millions of years ago but twenty-five thousand years back. Thompson stooped to pick up something beside a cornstalk, a fragment of white, soft bone.

"If I had a corn knife with me," he said, "I'd soon cut you out a little patch, so you could start digging."

He whittled off a number of stalks with a pocketknife, but I told him I would rather let it go for now and instead start early in the morning. That afternoon I rented a room in nearby Plymouth and Mrs. Hildebrand, my landlady, offered me the use of her roomy garage to store bones . . . if we got bones.

On my return next morning, Thompson lent me a corn knife to add to my tools. Walking back to the field, I could visualize the last great ice sheet retreating north, just yesterday in geological terms. It might not have been many miles away when post-glacial animals pushed up almost to the edge of the ice as vegetation worked its way north in the cool, wet climate. The creature buried here must be much like its cousin, the elephant. Whether it would prove to be one of the wooly mammoths, the big imperial mammoth, or the slightly more southern Columbia mammoth remained to be seen. The tooth Thompson had found was far away in the museum; I'd have to dig up the answer with other bones.

I went to work with the corn knife, enlarging Thompson's little clearing. Bone fragments were scattered through the muck, where the plow had evidently torn a bone below ground. When I had cleared a suitable area, I laid aside the corn knife and began to spade the soft ground gingerly. The shovel struck something that felt like rotten boards. In moments I turned up the fractured end of a bone. It was impossible to tell much about it until I had more of it exposed, but it was massive, and part of the mammoth without a doubt. I moved back and started an excavation in keeping with the creature's apparent size.

By five o'clock I had made quite a hole in the corn field. I found the bone to be part of the mammoth's skull and jaws, much the worse for its contact with a plow. Next morning, a foot or two lower down I uncovered, in addition to the remainder of the skull, part of an elephantine tusk. It looked to be three or four feet long, not large as mammoth tusks go, but a good-sized hunk of an old ivory nevertheless. I worked along it carefully. It was semi-rotten where it had lain near the surface, but the pointed end, driven deeper in the muck, was hard and well-preserved. Nearby I came upon the end of a large limb bone. All this in an Indiana corn field!

At noon Thompson came out to see how I was doing. Rather than being depressed by the havoc with his silage, he was curious about the creature that had wandered into his cornpatch long ago.

"It is evident," I told him, "that the big fellow bogged down here while this was still a big swamp, just after the end of the last Ice Age. In the moist peat, the bones have been protected from the air. While they haven't had time nor the proper matrix material to turn to stone, they have been preserved."

Before noon next day, several big ribs, part of another tusk, and the broad ends of several limb bones came to view under my shovel, brush and awl. I now had a hole some yards wide, two feet deep in places. Uncovering the bones went along much faster than working in the shales, clays and

Skull of *Machaeroprosopus*. Photo by Julius Kirschner. Courtesy Department of Library Services, American Museum of Natural History.

stony matrix I was generally used to. Early in the afternoon a few farmers dropped by to look into the pit.

"I know a fellow in town here who found some kind of skull in the bottom of a gravel pit," one of the neighbors said. "He has it in the barber shop. Looks like some kind of cow's skull, only it isn't."

I asked him about the horns, teeth, general details, and decided it was probably the skull of a musk-ox. The gravel pit was probably an old glacial dump or the place where a post-glacial stream, swollen by melting ice, had swept quantities of gravel. I took down the finder's name, planning to call on him as soon as I could.

"The musk-ox," I told the men, "once ranged as far south as Oklahoma. Now these cold-lovers have retreated to the arctic and southern Greenland. By the way, were there any other bones where the skull came from?"

"A bunch of 'em. But nobody thought they were worth anything; they all got thrown away."

I winced and went back to work. The rest of the musk-ox skeleton had probably been there, or a good part of it. Many of the bone records of the geologically recent past—the saber-tooth tiger, the musk-ox, and many other animals that are gone now—have had their records lost to us by insufficiency of curiosity.

That night Brown came in from Kansas by train. He had had a fine vacation, he said, and was ready to get into harness. I took him out to the dig and showed him he was more likely to get into mud. He was satisfied with what had been done but a bit disappointed in the specimen. The skull he pronounced pretty much of a loss but for the tusks. Grinning at my enthusiasm over my first mammoth, he climbed down in the hole with me and started shovelling.

Water was becoming a problem. I had the evening before struck quicksand below the muck, and this began to ooze like a spring. It became necessary to stop excavating bones from time to time and bail water. When we came upon the pelvis, the dorsal vertebrae, and a few ribs randomly placed, it was hard to bail around them. By this time the corn had been cut in the field about us, and I was able to bring the Buick up to the edge of the excavation. We couldn't use the Buick's water pump to advantage in our work, but it helped to turn the radio on loud and follow the world series ball game as we worked and splashed about. It was good to have Brown with me again, and in spite of the mud and the bailing, he puffed away at his old pipe contentedly as we worked.

That evening we drove to town to see the musk-ox skull in the barbershop window. The jaws and a few teeth were missing, but Brown was pleased with the specimen and thought it worth acquiring for the museum. The owner, however, was more acquisitive by nature than Thompson. To us, his asking price was out of line with its value. We left without the skull.

Next morning we looked out at depressing skies. For the first time in weeks, we were faced with threat of rain. When we got to the pit we decided it was time to start taking up bones . . . while they were still above the water from time to time. A few of the ribs and some of the smaller pieces could be removed from their positions without use of plaster, wrapped, and packed in newspaper. But the pelvis, big as a steamer trunk, the tusks, the crushed and beat-up skull, these must all be given plaster jackets. The skies grew darker and a few drops plopped into the water at our feet as we weighed and considered. We hurried to get jackets on the upper sides of our bare bones.

Then the downpour started. We threw pieces of canvas over our plastered bones. Water was now coming at us from above and below, and we had no raincoats. We were soon soaked to the skin and shivering. We got half a dozen ribs, a pair of tibia and fibula, and a number of foot bones wrapped in newspaper and piled into the back of the Buick. Somehow, working under the shelter of a piece of cold and drippy canvas, we managed to get some of the underlapping bandages on the pelvis. When we had slapped on the least we thought we could get by with, Brown looked at his watch.

"Two p.m.," he said. "I say call it a day."

The next day it rained, too. We expected to find our pit flooded, our mammoth all but afloat. So we were not depressed; things came up fully to expectations. Again the rain came on, but we gained on it. The tide at our feet down to wading level, we went on with the job.

In undermining one of the big limb bones, Barnum came upon a small section of water-logged tree limb, both ends of which plainly showed tooth-marks. It had been cut by a Pleistocene beaver, a creature very close to the beaver of our day. In another part of the pit we picked up a few hemlock cones, fallen from an earlier tree.

The rain changed now and then through the day from light to heavy showers and back again. We had hoped to empty the pit by nightfall, but when we climbed out of our hole we had to leave the great pelvis. Because of its weight it lay deeper than the other bones and was still far too wet to move. "We can only leave it until it stops raining," Brown decided.

The rains stopped during the night, but clouds still threatened us. We bailed out the hole again, but found the pelvis in no condition to move. Brown suggested we go for a drive while we waited for things to dry out a bit.

We drove west, naturally—away from home—feasting our wet eyes on new scenery exactly like the scenery we were leaving. Rich, black soil, the low places drained by ditching.

"Ditching has done a lot for us in this country," Brown noted. "All these low spots especially are good for prospecting for mammoths, and ditching is one way of turning them up for us."

"And in off seasons," he went on, "some of these folks probe the ground with steel rods; when they hit something they find out if it's bones or just a piece of wood."

"Sounds fascinating," I told him. "I'd like to come back and try that out, sometime. But not just at this time; I want a little while to forget our pit back there first."

At an unmarked country crossroad, we stopped at the nearest farmhouse for directions. From long force of habit, Barnum asked the farmer if he knew anyone who had any old bones.

"I got a skeleton in the woodshed," he replied. "A big one. Want a look at it?"

"I sure do," Brown answered.

The man led us back through the yard, threw open the door of the woodshed, and stepped aside. There ranged against the far wall and along one side were piles and piles of gleaming white bones, the major part of a mammoth skeleton, ranked up neatly like cordwood. They were in a perfect state of preservation and had long dried out nicely. They had been taken from the peat without need of plaster and were all unbroken and complete. Brown gave me a long look that spoke worlds.

"Where did you find this?" he asked the farmer.

"Oh, I uncovered it out here when we were draining a field."

We looked it over carefully. Here was a representative skeleton of the locality at its very best. Brown told of our purpose in coming to Indiana and dropped a hint that the museum might be interested in these bones. The farmer named a figure that seemed pretty steep for one animal, times being what they were. We left the woodshed and went on to continue our tour. It reminded us somewhat of the musk-ox skull incident.

"These people set quite a store by their fossils. After all, bones are bones. We might as well take ours and go home."

We went back to our cheerless pit in the corn-field. The flowing spring had again raised the water level to the bottom of the pelvis, but the jacket seemed solid enough to chance handling it at last. After bailing about a hundred buckets of water from the hole, we undermined the specimen and got it to the bank. Luckily it just fit into the Buick's trunk carrier. We refilled the excavation we had made, gathered our tools, and went to the house to thank Mr. Thompson for his part in making available to the museum what, in happier circumstances, would have made an excellent display.

In town we retired to the local lumber yard to build crates. Again soggy rain fell from sodden skies. The bones from the garage were piled in the back of the Buick and brought around to the pelvis. As we worked in the open, the packing straw got wet. I wondered how the soggy mass would influence the man who finally opened the crates at the Museum. I hoped I'd never find out. When the shipment was finally billed out at the freight office and a trucker came to gather them up, we were filled with a strong sense of relief.

Mrs. Brown arrived on the evening train from New York. Next morning the three of us drove east from Plymouth. A bright sun laughed, perhaps jeeringly, in our faces. Mrs. Brown commented on the healthy tan on her husband's face; he hadn't faded appreciably in the dreary days in the pit.

She laughed teasingly, "How does Brownie manage to look so fit, when he's so long away from me?"

"Don't you tell her; don't you dare ever tell her!" Brown warned me.

Restoration of Jefferson's mammoth (formerly known as the Columbian mammoth). Most of the mammoth fossils in Indiana probably represent this species. Painting by Charles R. Knight. Courtesy Department of Library Services, American Museum of Natural History.

17

THE LONG TABLE in Brown's office was littered with maps and aerial photographs. He had just held a press conference. I floated by the open door, treading lightly on clouds; winter had come and gone; the air was charged with plans for a new expedition. We were to head out in a few days. The press conference was a prelude to the fanfare of publicity that spoke of a full-fledged party.

Next morning the story appeared in the papers. A three-toed footprint of giant size had been found on the roof of a coal mine, the Chesterfield, in Utah. The creature that made it, the paper said, was pronounced by Brown to be "unknown to science." No skeletal trace of this giant had ever been found. But before this expedition was done, before the season was over, we hoped to bring back some of the bones of this mystery dinosaur.

I read the accounts over and over for the exuberant feeling they gave me. The much-publicized track was on exhibition in the Seventy-Seventh Street foyer, thoroughly photographed at the time of the press conference, and the pictures accompanied the articles.

I was downstairs packing when the phone rang. Louis Monaco, of Brown's office, was on the line.

"Sorry to interrupt, R.T.," he said. "Hoped I would catch you there. Can you be in the foyer at two o'clock? Matter of more pics, this time for Sinclair. Brown will be there. Bob Chaffee. Gil Stucker. Erich Schlaikjer, if we can find him. Don't forget."

Schlaikjer, of Brooklyn College, had been in and out of the museum lately, planning to join the trip west. Gil Stucker, a new man, was going along. Bob Chaffee, my classmate as well as lab companion, was also to go along. Don Guadagni and Ted Lewis were also to accompany us, but so far neither had shown up in New York. The Sinclair Oil people had again kicked in generously toward an expedition, and these pictures were for their own special use.

At the appointed hour I found everyone, including Brown, rallied around the footprint. Chaffee and I were given a yardstick, and we went through the motions of measuring the print, while Brown and Stucker pretended to look on. Thirty-two inches across, black as night with the print of coal. The label described it as the "biggest footprint on record." There were no signs of claws on it, but the great pads gave it some affinity with the *Iguanodon*, a bipedal second cousin, in a way, of the *Trachodon*. Brown estimated the creature may have stood twenty-five or thirty feet high. We were starting out in search of a spectacular animal indeed.

Our first destination was Rock Springs, Wyoming. Here were outcrops of the same formation as that in which the giant track had been found. This formation, the Mesaverde, was Cretaceous in age, dating eighty million years back, but no dinosaurs had ever been collected in this particular horizon. It was an extensive series of sandstones, clays and coal seams aggregating to a thickness of two thousand feet in places. It covered hundreds of square miles in southern Wyoming, western Colorado and eastern Utah. The bone outcrop near Rock Springs that Brown had located in 1934 on his aerial survey and which he had so long hoped to work might well prove to hold the bones of our mystery dinosaur.

It was planned that I would set off for Rock Springs with the advance party and establish camp. Brown hoped to follow by the end of the week, and Schlaikjer and Lewis would follow as their respective colleges closed.

"When you get to Rock Springs," Brown advised me, "look up Dr. Lauzer at the local hospital. He was mayor of the town once, and is a man of some influence there. He'll take you out and show you the bone outcrop. If there's anything you need or want, he's your man."

Brown continued, "Then there's the steam shovel. The Union Pacific writes me they can deliver it on the siding near the foot of Number Five Hill within the next few days."

This was good news. Brown had been negotiating for the use of a machine to strip away the overburden, somewhat similar to what we used at Howe Quarry. That the Union Pacific had been generous enough to donate a power shovel to our cause would mean elimination of a tremendous amount of hand labor.

"If we find the Big Fellow, we'll need every resource we've got," Brown added. "I'll be with you shortly after you get set up. Good luck!"

We set off next morning, Stucker in his own car, carrying Chaffee and his baggage, Guadagni with me in the Buick. We hit the Midwest with a record June heat wave, but by the time we got to Rapid City, South Dakota, we wished we had kept some of it with us. We put up the Buick's side curtains and turned on the heater. Next morning, in Gillette, Wyoming, Don and I looked out the window at four inches of soft snow. Don dug out his heavy coat. "But for the sweat we were in earlier this week," he said, "I might think this is Siberia."

We went up to Billings, Montana, to pick up the Ford truck stored there since the Howe Quarry job and detoured for a brief visit to the Patons. No snow, but we still had winter weather in the summertime.

We reached Rock Springs from the east, and looked a bit critically on the bare, spread-out little coal town set in gaunt, treeless surroundings. We must learn to like it; it was to be our home for a good bit of time. North of town I recognized Number Five Hill from Brown's photographs, a grey ridge so lacking in individuality it seemed appropriate it should be numbered instead of named. We established ourselves in a tourist camp at the base of the hill, about a mile from town, and I set out to look up Dr. Lauzer, head of the local hospital and one-time town mayor.

He volunteered to drive us to the site of Brown's find, and we were soon on a steep and tortuous little trail that led us past grey sandstone walls to the top of Number Five Hill. "Hardly anyone ever comes this way, except jackrabbit hunters," Lauzer told us.

At the top of the long bench-like hill, the trail turned south. From Brown's aerial photographs I recognized, in the thick-bedded brownish-red sandstone alternating with bands of brown and purple clays, a part of the upper Mesaverde. The formation rose sharply, the sandstones resisting erosion and forming broken ridges above the shales. For a mile the rough trail wandered south among these rough sequestra. A thin scattering of sagebrush grew along the wayside. We stopped at a spot Lauzer seemed to recognize. It proved to be the bone outcrop, a few small and scattered fragments showing, nothing indicative of what might lie under the large ledge of overburden.

"Brown didn't attempt to uncover bones," Lauzer said. "So open, so close to town, he didn't want to risk it."

We went back to Rock Springs. Number Five Hill looked good for bones but uninviting as a campsite. But if we were to open a quarry there, we couldn't stay in town and leave it unattended.

Finding the boys, I led us all back to the top of the hill. The ground was high and dry. Not a tree for ten miles. Exposed to the winds from every direction but underneath. While we moaned and groaned, Chaffee and Stucker set up the cook tent. I set up Brown's famed India tent, not unfurled since Howe Quarry days. Don Guadagni set up a green umbrella tent barely bigger than he was.

All through the night a cold wind from the south flapped the tent, drifting dust in on us under the flaps, a fine and sooty dust. It covered our clothes, crept into our beds. For breakfast it blew into our food, giving us a low fiber diet, high in mineral content. Vitamins were extra.

While the rest went about setting up another tent, I got out tools and began to explore the most likely section of the meager bone outcrop. I dug into the shale randomly and exposed the end of a rib. When it began to sprinkle, I covered this up and went to help the boys.

At the Howard Cafe in Rock Springs, the waitress eyed us curiously as Don and I gave our orders. "Are you boys from New York?" she asked.

When we confessed we were, she added, "Well, then; it's you in the papers. You've sure started something around these parts."

I bought a paper to find what had brought this on. "First of the Dinosaur Hunters Arrive," the headline blazed. The *Rock Springs Rocket*, undoubtedly with the aid of Dr. Lauzer, had made our expedition the feature of the day, including the pending arrival of the steam shovel. Actually, what the

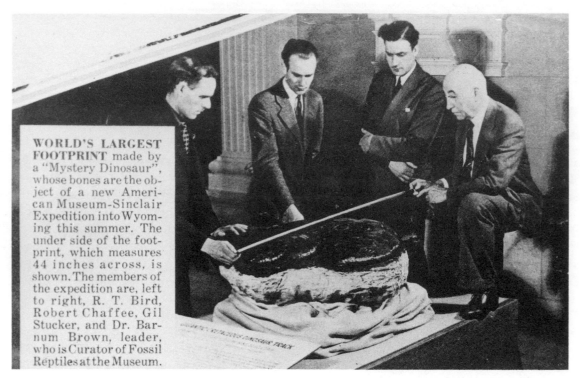

WORLD'S LARGEST FOOTPRINT made by a "Mystery Dinosaur", whose bones are the object of a new American Museum-Sinclair Expedition into Wyoming this summer. The under side of the footprint, which measures 44 inches across, is shown. The members of the expedition are, left to right, R. T. Bird, Robert Chaffee, Gil Stucker, and Dr. Barnum Brown, leader, who is Curator of Fossil Reptiles at the Museum.

Bird, Robert Chaffee, Gil Stucker and Brown contemplating the huge ornithopod dinosaur footprint from the Chesterfield Mine, Sego, Utah. *New York Times* photo.

railroad people had contributed was a drag line, more suited to our purpose than a shovel. As the paper reported the event, the big machine was already on its way, covering the ground in part on a mobile track of railroad ties. The press made something of the fact that this was the first time such a machine was to be used in probing for dinosaurs. In the past, most work done by archaeologists and us paleo people was wrought with awls and whisk brooms, with simple picks and shovels serving for the heavy work. The wording of the story led the reader to wonder how dinosaur-hunting had ever been done without the aid of equipment of comparable size. It closed with a word picture of the giant we had come to find.

"We'd better find a dinosaur now," Don remarked, "or we're all gonna be sorry."

About three o'clock, I went to meet the afternoon train; Brown was on it, breaking precedent by being on schedule. He was promptly surrounded by pencil-bearing reporters but escaped as soon as possible, as anxious as the rest of us to see the big job off to a big start. We squeaked by the big shovel, still wallowing its way up to the site, its great bucket waving from side to side on its long neck, for all the world like one of the great beasts it had come to hunt. The operator called out, "We'll be ready to dig in, first thing in the morning."

Brown's arrival was featured, of course, in the morning paper. The town hummed with anticipation of great events; by the time we got back from breakfast, several carloads of sightseers had crawled up the trail to our camp. Brown told the operator to start up at once, explaining he wished overburden removed down to a foot or less of the anticipated bones. The rest of us gathered around, awls and picks and prospecting shovels in hand, waiting for the big machine to strip down to the bones.

We watched the big scoop drop down, settling its sharp teeth into the shale, bite deep, and draw back, ripping a yard or more from the earth. The bucket rose and swung aside, dumping its load some sixty feet away.

Brown's eyes shone. For minutes nobody did a thing but watch the big-mouthed shovel bite, swing, and spit 'er out. This was bone-digging like it should be.

The operator cut his motor, dropped the bucket, and shouted down to Brown. "How'm I doin'?"

"You're looking beautiful. Just keep her up. I'll let you know if you cut too close to the bone."

The operator worked on, until the cavity looked like a highway cut or preparations for a parking lot. Gil Stucker was assigned to watch the dump for bone fragments. Our early visitors arranged themselves on convenient sequestra, and newcomers, arriving in a steady stream, placed themselves as they could. Dr. Lauzer arrived with Mr. McAuliff, an official of the Union Pacific Coal Company, to meet with Brown and observe the digging for a bit.

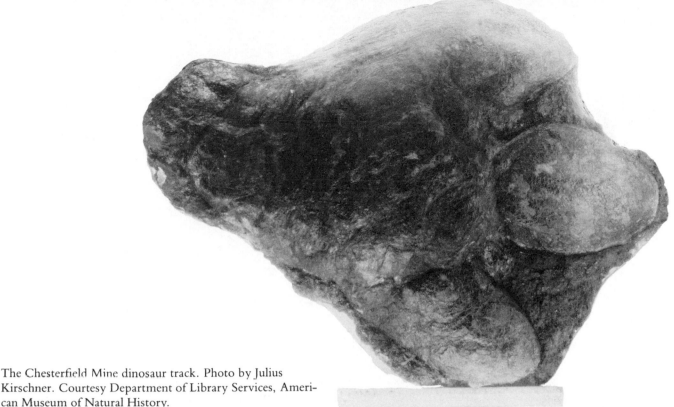

The Chesterfield Mine dinosaur track. Photo by Julius Kirschner. Courtesy Department of Library Services, American Museum of Natural History.

When the bucket had moved along far enough for us to go to work safely, Don and Bob and I moved in on the outcrop, spread about, and went to work with eyeballs and hand tools. The matrix was soft enough to work easily with awls and small picks. Fragments at the bone level ran into the matrix here and there, but were all broken fragments. I hunted out the end of the rib I had noted on the first day and dug it out. The bone was firm, though the head was missing. Brown took notice of it.

"Not much of a start, but it's a start, at least," he commented.

At noon we took turns going to the Howard Cafe, as before. We had to buck the stream of steady sightseers on the way down the trail. The trail was beaten until the ruts were filled with floury dust that swirled about us more indifferently than the visitors.

Our gallery at the dig continued to grow . . . and to grow impatient. Our show wasn't up to expectation.

"Where are all the bones?"

"Is that all you've got, so far?"

"Will there be more, over where the shovel's digging now?"

Our part of the script was brief: "We don't know."

We moved the shovel and bit off another cubic yard of landscape just a little farther down the ridge. Brown watched closely for bone sign. Don and I pushed the ground work as close to the shovel as we dared. Gil, on the dump, watched closely. The shovel continued biting its way through the brown sandstone. Gil called over the shovel noise and the rattle of heavy rocks.

"This rock's got fossil plants in it!"

Brown signalled to halt the operation, and we all gathered about Gil's display. Right through the brown stone was a mass of clear leaf impressions, mostly deciduous, diagnostic of Cretaceous time, when deciduous trees were first established. No plant material was present, only a thin carbon film. A clear impression of a small palm leaf mixed in with the other impressions. We moved the best pieces back to where they wouldn't be covered. "If nothing else," Brown commented, "we can start a nice plant collection."

When the big machine shut down in the afternoon for refueling, it was too late to bother adding more to the day. We all scurried around, digging by hand and with hand shovels through the reddish-brown and brownish-green matrix . . . well, no; it wasn't matrix, really. It was just dirt; to be matrix it has to have something in it. It was just dirt, empty dirt. Our audience drifted away as silently as our high hopes.

At bedtime, the broken rib I had dug out—the find of the day—was set up on its pillar of matrix behind the tents. Our lean emblem of hope. Our pledge to our audience.

Next morning it was gone.

18

BARNUM BROWN led a strange procession, so strange, so special, I wanted the worst way to have it preserved for posterity on film. Unfortunately, there was no camera loaded, and I couldn't lay hands on a film. Brown walking out the long ridge of Number Five Hill, a little distance from our recent and disappointing excavation. Gil Stucker and Bob Chaffee in his wake. Big Don Guadagni, our Patagonian giant, coming behind. Then the shovel, its great boom swaying, more like a lumbering dinosaur, a lost and disappointed iron creature in search of its ancient meaty kin. One by one they all topped the small rise and disappeared over the brow, all but the shovel's steel boom.

Hurrying to overtake them, I reached the top in time to see Brown and the boys bending over what appeared from a distance to be fragmented bones scattered about the surface. There was no indication they might have leached out from the hillside. Brown suggested, however, that the shovel might cut into what we hoped might be the bone layer.

The operator dug into the hill with a repeated monotony of movement. Bob and Gil watched the dump carefully. Don and I stayed as near the grinding blade as we dared. Suddenly Gil cried out, and Brown signalled the operator to halt. The rumble of the shovel died, and we all hurried to the dump to see for ourselves what Gil had noted. Gil was prodding into the back of a ten-foot pile with the point of his prospecting pick. A bone fragment fell out with a lump of clay, followed by a shower of tiny bits of bone. What the shovel had ripped into was broken a million ways. We tried to ascertain what the bone had been, but it was too badly fragmented. Brown stepped into the pit, examining the spot where the shovel had taken out its last load. Nothing to be seen.

"Well, anyway," Brown said, "we know there's been a bone here . . . that's something." He turned to the shovel operator. "Now, then, if you'll move just a little farther over, we'll see what we find."

While the shovel trundled to a point where it could bite into a new area, we all crowded into the hole the bone had come from, each eager to explore the matrix. In a few minutes, Don's awl struck bone, and he and Chaffee soon exposed it to view, a big flat bone, much bigger than my stolen rare rib. It was fairly complete and seemed a part of a skull of sorts. Brown pronounced it a part of a ceratopsian skull. There weren't enough pieces to say accurately which one; it was a large and varied family. We could each of us visualize any one of these lumbering four-footed herbivores waddling through an ancient mixed forest in company with our mystery dinosaur. Each of us was free to fit out our visualization with horn nose-pieces and any of a variety of neck-and-shoulder guards to suit each fancy. We were without enough clear evidence to challenge anyone's vision. While we were variously thus engaged, a chill wind whipped a few flakes of snow about.

"Just think, it was tropical here once," Don commented. "We should have come earlier."

At the end of a week that largely resembled our opening day, we had thoroughly prospected the three-mile length of Number Five Hill with unmitigated lack of success. We gave the shovel back to the railroad and would have given away our meager supply of bone scraps, if there had been enough to go around. Recalling that first rib—still one of our best finds—we were confident of only the slowly flagging interest of our still persistent guests. We moved camp seven miles to a spot Lauzer and Brown had scouted out before, on the west side of Eden Valley on the flank of White Mountain. There were a few cedars and other desert growth about, not enough for much protection from the wind, but a place where we felt less naked to the elements. And we installed Maw Denniger, of Howe Quarry fame, who had been waiting in Rock Springs for us to settle down. With her great camp cuisine, we ate better. We didn't do better, but doing isn't everything; eating is important too.

Despite our failure to turn up giants and solve mysteries, the shovel experience was enlightening and changed the future of our sort of work a good bit. We hadn't found a thing, but in vastly less time than would have been possible by hand methods, we had found out a lot that wasn't so . . . which is every bit as important anywhere. In fact, Brown spoke enthusiastically of perhaps one day having a small power shovel or drag line of our very own.

The situation our party now faced was unique in my experience and, I believe, in Brown's. For three years we had looked forward to an expedition at Rock Springs. The Mesaverde was a new field and, until our work in it, a promising field. In striking contrast to Howe Quarry, where we had been utterly overwhelmed by a plethora of material, here we were indeed underwhelmed by a paucity of anything. We had two things going for us: there was still a tremendous area of virgin country to explore, and the summer was still young.

We had a third thing going for us, too, really: Erich Schlaikjer. He came on the scene late, because of his Brooklyn College commitment; he hadn't shared in the overwhelming disappointments of Number Five Hill. He brought us a lot of "Let's Go" spirit at a time when we all could use it. He didn't know the Mesaverde formation was for the birds rather than for their progenitors. He whooped with joy when he came upon a stone with a bone sticking luringly out of the cold and disappointing ground. "Who says there are no bones in the Mesaverde?" he asked everybody. Everybody—that's us—picked up our tools and went on with the job.

Sunday morning Brown and the rest of us were ready for a day off, ready to do something desperate: nothing. Erich, fresh and full of beans, wanted to hurry out and look into some of the Mesaverde he had seen from the train. From the train window the Baxter Basin, between Rawlins and Rock Springs, looked ever so good to him. "Come on, R.T., let's drive out and look it over," he urged.

We climbed into the Buick and headed east on U.S. 30. The highway passed under a sheer yellow wall, the Golden Wall, the basic sandstone of the Mesaverde. Beyond, the road ran straight east across the Baxter Basin. The roadside soil was gumbo clay, soft and grey. The Cody shales, Brown had told me, marine in origin, went back to the days when this was a great Cretaceous sea, before Mesaverde time. Dinosaurs, lots of dinosaurs, hung around the water a lot, but none of them conquered the deep. We were about an hour driving through this; then we rose again through a continuation of the Golden Wall, then through the coal-bearing formations. Bitter Creek, nearly dry, ran alongside us for a way, then turned south and left us, running under the high escarpment of the Ericson formation. High over our heads, perched on a crag where it could be seen from the far side of anywhere, was a steel airplane beacon, at a spot marked on our maps as Point of Rocks. We climbed up, up and up, until we reached the level on which the beacon stood and decided to get out of the Buick and stretch our legs with a bit of bone prospecting. We separated, each following a bench that made good footing and was located at what we hoped was a promising geological level.

Happily I soon came upon a bone fragment, no more than a few inches long. I pounced upon it eagerly, as I was able to identify it; it was not merely a bone fragment, it was part of the dentary of a ceratopsian jaw. I was quite sure as a result of weeks spent in the museum working on the *Styracosaurus*. I had picked this up in a gully; possibly the rest of the great jaw might be buried above me. Equally possible, all the rest of the great skull might long since have weathered out and gone downgrade below the point where I found the bone. This is the way the cards are cut. I scouted the downgrade, of course, looking at the dark side first. I found no other sign of bone and worked my way back toward Erich and the Buick.

I found Erich hunkered down happily over a bit of what seemed to be an ilium projecting from the bank. "Who says there are no dino skeletons in the Mesaverde?" Erich asked again.

I hoped he was right; again, the question was whether this was the butt end of a buried dinosaur. Or was it the last bone left from an old fellow who had weathered away and gone down the slope ages ago? Only time and good digging would tell. We marked the spot for future reference, as it was time to get back with the party.

After our reconnaissance of Point of Rocks, it was decided prospecting the west side of Baxter Basin should become the order of the day. With three cars and an experienced staff now at our disposal, we hoped to recoup the losses which Lady Luck had thrown us so far. We could split up and spread our prospecting, or we could all pull together, as the occasion demanded. Late in June Brown ruled in favor of reinspection of a site looked over briefly near Winton, near an area known as Boar's Tusk. As this was beyond paved highway, we would have to pound and grind our out our own trails. Boar's Tusk was a great monolith rising sharply from the Eden Valley floor, the hard lava core of a volcano whose softer sides had long since eroded away. Near Boar's Tusk a lean and rugged trail took us eastward toward a distant line of ridges that became our objective. A difficult trail, too steep or too sandy, by turns. In two miles, we had the Buick's radiator boiling furiously and demanding surcease. Fortunately for the old car, the trail ended there in a dry wash between two ridges, encouraging us to scout out the banks. The spot was marked on our map as Long Canyon. The day was as hot and blistering as the Buick, and in an hour we were back at the car, empty-handed and sweaty-browed. So we wiped off the sweat by hand and checked out the other side of the ridge. Then another little ridge, just a bit farther on.

We reached the crest of the ridge. Ahead, an open spot of weathered shale stretched out just under a sandstone outcrop. The sandstone was thicker just at this point and rose slightly higher as though, prior to its deposition, running waters had created a channel in the underlying muds. I had seen many such outcrops, and could always picture a skeleton caught and buried at the channel side. I saw a line of fragmentary bones scattered around the surface, like bits of gravel. The sunlight cast black little shadows under their sharp edges.

"Hey, Erich! Here's something!"

He nodded when he saw the projecting bits of bone, not yet impressed by what might lie below the overburden. We dug carefully along the line of scattered pieces. We were excavating what turned out to be a line of very large ribs, evenly spaced, as if they might still be attached to vertebrae.

"Looks good, real good." Erich said. "R.T., maybe we've got something."

Having been through a bad time that Erich had missed, I was afraid to believe. But when the boys came up with the water bag and the lunches, we all tackled the overburden. In one place we uncovered what seemed to be the end of a giant femur. Farther along, articulated caudal vertebrae. There seemed little doubt a huge dinosaur, more or less complete, lay under the heavy shale. But at four o'clock we had to leave the ridge and head back toward the main road.

We got back to camp just ahead of a dust storm of very fine dust. The coal black cloud warned us of its coming before we reached camp. It blotted out Aspen Mountain, then Rock Springs, with its black curtain. In the faint lazy stirring of the hot air, we hurried about checking tent stakes, battening down in general for what we knew was about to hit. Then the still hot air suddenly drew a deep breath, gasped, and was swept away as abruptly as cannon shot when the dust hit us, testing our tents and tent stakes.

Brown and Gil arrived in the middle of the storm, coming into the whipping tent as Maw Denniger was trying to prepare and protect supper. "That's enough of that quarry; we won't go there anymore," Brown announced. "And how did you boys do?" When we told him how we did, he was more elated than he had been in days.

"R.T.," Brown said, when we had kicked the good word around, "this means you and I can get away, perhaps tomorrow even, for an inspection trip to Cedaredge. I've sort of held back on it because of the troubles we've had. But now the rest of the boys can go to work on the dinosaur you and Erich have run to earth while we check this Cedaredge thing out. If it comes up to expectations away better than this Rock Springs deal, we may finish out the summer more or less even up after all, maybe even ahead of even."

"So what's Cedaredge?" I inquired.

"Well, a Charles G. States, a coal mine owner in Cedaredge, Colorado, says he has some giant dinosaur tracks walking across the ceiling of one of his mines. A lot of 'em. He says this fellow took fifteen-foot steps. Mr. States thinks one of us scientifical fellers ought to have a look at them. And he may be right. They're in Mesaverde coal, by the way."

"You mean this is sort of a . . . could we call it a . . . maybe a mystery dinosaur like the one who made the prints in the Utah mine?" I asked.

Brown chuckled, a bit wryly. "States says he's even found some bones. And, apparently the fellow didn't die around here."

"You mean, Boss, it could be the same one?"

"Fifteen foot steps? He could have gone over there for lunch. And got ptomaine poisoning. And maybe died there. In fact, I bet he did." Brown took another sandy puff. "Served him right. Damn him!"

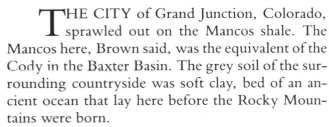

19

THE CITY of Grand Junction, Colorado, sprawled out on the Mancos shale. The Mancos here, Brown said, was the equivalent of the Cody in the Baxter Basin. The grey soil of the surrounding countryside was soft clay, bed of an ancient ocean that lay here before the Rocky Mountains were born.

We arrived late in the day and left in the early morning for Cedaredge. A gravel road led east. High mesas. Towering walls with a familiar look; the sandstones and coal-bearing measures of the Mesaverde. The eroding Mancos rose to meet them like rough waves beating against cliffs. All blue with the haze of distance.

At the village of Cedaredge we learned the States Coal Mine lay a mile or so to the north, at the foot of Grand Mesa. Our road climbed the last of the rising Mancos to enter a little valley, glaring green with alfalfa fields, soft green with rows of aspen along the fences. It passed under a cliff of brownish-yellow sandstone seventy-five to a hundred feet high. Brown studied it intently from his side of the Buick.

"That's the Rollins Sandstone," he told me, "Probably the equivalent of the Ericson in Rock Springs. About the Golden Wall . . ."

A tall coal tipple interrupted the correlation. Tipple and buildings were weather-beaten, the tired grey of old barns, contrasting with the bright gleam of freshly mined coal in a chute beside the tipple track. A lazy curl of blue smoke, acrid smoke, rose from a nearby burning slag pile. The road passed between a small bunkhouse and an office building, terminating in a turn-around under the chute. We parked the Buick beside the bunkhouse and inquired about Mr. States from a man standing in the boiler room door.

"He'll be down from his house right away," he told us. "He seen you coming."

The man who hurried down the path past the tipple was perhaps sixty-five, short of stature, mild of manner and speech. He had been expecting Brown and was delighted to see us. He began telling us at once about the tracks in his mine; their discovery had been an exciting event in his life. He led us into the office nearby, picked out a carbide light, and began tinkering with its mechanism.

"I know you'll want to go underground for a good look right away," he said.

We slipped on old coats, and I lighted up a gasoline lantern we had brought from Rock Springs. We followed States up a little narrow-gauge railway leading to the mine entrance, then single-file into a timbered tunnel, where a cool breeze with a coal smell fanned us.

"We're really headed for another mine, the States-Hall," States told us. "It connects with this one. Easier to walk through than to drive around. The tracks are over there."

At first, coming out of the bright day, our yellow lights seemed feeble indeed, but as we got deeper in the dark and our eyes became accustomed, our lights seemed to brighten greatly and to strike bright gleams from the black coal walls. We were beyond timbers now. The overhead was a smooth sandstone, medium fine-grained, a stone page in the Mesaverde sea-world where sunlight hadn't shone for eighty million years. The coal seam was eight feet thick. We were constantly passing, or entering, tunnels along the way. States led us right, left, right, right, left, unhesitatingly. Here and there along the way, miners worked busily.

We passed the last of the working miners and came into broad and tremendously high rooms. "The coal seam is fourteen feet here," States told us. "Solid. In some places it is cut by a sandstone stringer, but here . . ."

I computed in my head a little. A swamp a

One of the dinosaur tracks in the roof of the States Mine. Photo by Bird.

hundred feet thick with dead plant material had lain here ages to form this great bed of coal.

Our own little noises heightened the oppressive vastness of the silence. "We haven't taken coal from here in years," States told us. "The workings over there on our right are abandoned."

We were going downgrade steadily and noticed a growing trickle of water in a gutter along the edge. A compressed air water pump chugged along at its job. At the end of several hundred feet of straight passage we came out into an enormous room with a rail track of broader gauge. States raised his light and flashed it along the sandstone ceiling.

The glancing light of his carbide lamp caught a great welt, highlighted and shadowed in places and by turns. As he shifted his light about we could see what it had to show: three tremendous toes branching out from a broad foot. Nearly a yard across. No sign of claws. The mystery dinosaur in the footprints, if not in the flesh. We were looking up at the bottom of a footprint.

"Big fellow," States said. "Fifteen foot to a step . . . but I told you already."

I looked ahead, in the direction the toes pointed. There in the shadows another great welt stood out. States took out a steel tape and insisted on proving what he had said. "See? Fifteen feet and better."

Brown turned to me. "States tells me the boys will have another track cleaned up in a day or so. What a thing, to get a slab carrying a complete stride!" Turning to States, he asked, "How thick is this sandstone stringer?"

"Only a foot, maybe two. And it's only a parting; above is the rest of fourteen feet of coal."

"We'd have to mine the coal above this, of course," Brown pointed out, "to make room for lowering tackle. The slab would have to be taken out in no less than three sections."

I was exuberant to think he would attempt such a task. Here indeed would be a spectacular exhibit. Mr. States was even more thrilled with the idea that his tracks might become a part of the famed American Museum of Natural History of New York. Grand old Mr. States would never be concerned about going to heaven; heaven would have come to him.

"People would never believe the stride, unless we bring back all the rock between," Brown added. "Maybe eight . . . maybe ten ton."

We moved out into the main entry and stepped aside for a string of cable-drawn coal cars to rumble by.

"Now," Brown inquired, "how do we get out of this place?"

States led us up a steep rise toward a tiny speck of daylight, and in a minute or so we emerged in the glaring day.

"Now about those bones you found . . ." Brown suggested.

"The best one is in the office. And I didn't find 'em; they came from Green Valley Mine, just across the valley."

We followed him to the office where he showed us a humerus about two feet long. It was a beautiful stone bone, as fine as I've ever seen. But it could have been no part of our mystery dinosaur. Brown pronounced it a part, in all probability, of a *Trachodon*, a duckbill dinosaur. Possibly a distant cousin of the beast we saw, but that's the best it could be.

The tipple of Green Valley Mine was in sight, and we went to visit it as Mr. States had other business for the afternoon. Green Valley was not a walk-in mine; it was a deep shaft, entered by a cable car which dropped us through the dark down a forty-five degree grade, a prosaic thing to miners who do it every day but for a first-timer, assuredly the thrill of a lifetime. Hutchins, the mine boss we were referred to by States, led us through passages exactly like those of the States-Hall Mine. He showed us the very depression from which the humerus had come, and Brown chipped from it a small splinter of the bone that had remained in the matrix, part of one of the condyles. We looked carefully across much of the ceiling. The lone bone had undoubtedly been a drifter, an isolated part of a skeleton washed here eighty million years ago by flood, long after the creature who had used the bone had died.

"Here and there we found a few other bones, stuck in the ceiling like this," Hutchins told us. "But they made no sense together."

We visited other mines in the area, saw a few other bones and a few other tracks, shot pictures of the big tracks on Mr. States's ceiling, even detoured to look over the battered remains of a mosasaur, a great sea serpent, removed by a miner with a pick and shovel. Then Brown said it was time to head back for Rock Springs.

20

BACK IN ROCK SPRINGS I asked how the crew was making out on the Long Canyon dinosaur. Bob shook his head. "You and Erich sure started something, finding that skeleton," he said.

Erich ducked into his tent and emerged with some very large teeth, long and slender with roughened crowns, wrapped in paper. "*Trachodon* teeth," Brown pronounced them, "but bigger than I've ever seen." The boys had taken up several blocks of bones but had removed none of them from the quarry.

"We don't know how to get the truck up the hill," Don confessed. "But one thing sure: nobody else is going to get up there, either. Our bones are safe; they won't be bothered there from now 'til Doomsday."

Erich reported finding another ceratopsian skull, about five feet long, near Rock Springs airport just across the valley. Guadagni had located, not far from Long Canyon, an outcrop of fossil shells and plants, adding materially to our Mesaverde collection. Ted Lewis, while we had been away, had rejoined the group. Mrs. Brown arrived, prepared to step into her old role as Barnum's secretary. All in all, it was like old times at Howe Quarry.

"Well," Mrs. Brown remarked, "what with your skeleton, and odds and ends of skulls and shells and plants, plus your big tracks, you boys and the professor certainly won't have to go back to New York crest-fallen and empty-handed, eh?"

Next day Brown and I climbed Long Canyon quarry with the rest of the boys. Dr. Lauzer and his wife came along to see the first of our complete Mesaverde dinosaurs.

At the hilltop we looked down into the excavation—not bad for a job of a few good old boys in a few weeks, all dug with hand tools; it ran back

eight to ten feet into the bank along the outcrop, exposing an area about ten square yards in size. Added to this, they had done a whale of a job with the skeleton.

Partly bare and partly already well done up neatly in plaster jackets, the sprawled outline of the monstrous *Trachodon* was plainly discernible. Here was no great lumbering sauropod, but a two-legged duckbill who could have stood up and looked them in the eye. Brown turned to Erich.

"The next thing: how do we get the truck up here? Do you think it can be done?"

Erich laughed. "Consider it done . . . pretty soon. I just didn't want to show off without an audience."

The truck sat at the end of a trail already broken to a point where it seemed the next thing was to break the truck itself. Ahead the ground was littered with weathered stone slabs, sort of shingled with patches of shingles put down rightly, others upside down.

Erich got into the truck and drove it side-saddle to the slope, angling up at eight or ten degrees. He crawled along until the wheels began to do nothing but spin. Then he threw it into reverse, and as the truck got momentum, pulled on the wheel and cut back up at a new tangent. Now and then, it looked as if the truck would roll over. He got support from the others, who stood on the fender on the uphill side, to keep him from rolling over. At the end of a boring series of this exciting maneuver, he parked beside the pit. Truly the trucking industry lost a Paul Bunyan of the trade when Erich Schlaikjer opted for a dull doctorate instead of sticking to a line in which he showed such aptitude. This, however, was the easy part of getting our *Trachodon* exhumed and back into our own century.

By mid-July this part of the overall job was

well enough in hand for Brown to consider dividing his forces in the interest of more conquest.

"I'll go with Bob and work at Long Canyon," he said. "Ted and Gil and Don might go back to the skull at the airport. And you two might as well scout out the area north of Dines Road."

Erich and I parked the old Buick in a little draw and started off on foot, in a blazing noon sun, up a little ridge that looked promising. We had shuffled about for something like an hour when we raised our heads, sniffed the breeze, and looked to the north. The blackest of thunderheads loomed, threatening rain. "Better we go back to town until this blows over," Erich said.

We drove in to Rock Springs, to hole up awhile at the Howard Cafe. The big storm turned eastward. We watched the rain blot out the far side of Baxter Basin, but the sun was still shining brightly over Dines Road. So we went back to the field, a bit embarrassed. We parked the Buick again in another little draw that looked good and went to work poking at the grey and purple shales. Almost at once I picked up a few bone fragments. An isolated humerus. Nothing of value, but encouraging. A little further on, a few more outcrops of isolated limb bones. Erich came over to look at what I had. "There's more material lying around loose here," he said, "than we've seen in the Mesaverde."

Thunder growled at us distantly. The black clouds of morning were still in sight but only hanging around aimlessly. We gave them little attention, as we had come upon three disarticulated vertebrae . . . broken and scattered, but a promise. We bent together over the largest, looking for possible continuation of a bone layer. Thunder growled at us again, closer, but . . .

The shock waves of the next crashing thunder shook the hills. Us, too. "R.T., did you hear that?" Erich asked, in hushed tones . . . particularly hushed on the heels of such a cannonade.

A thunderhead had piled up on the north, all over White Mountain, clear across the Boar's Tusk country. The sky beneath the pillar of piled black clouds was black, blacker than midnight. We watched it move down Eden Valley in fascination. If it came on south . . .

It came to me we had left the Buick in a canyon that cut across the Mesaverde to drain part of Baxter Basin. The deluge headed our way would fill the narrow throat between its walls in an hour, carrying the Buick off like an empty matchbox. And it was my day to drive; my responsibility.

"My tired feet, if we lose the car," Erich added. "Let's roll it!" We both leaped up from our old bones.

"No use for both of us to go for the Buick. I'll cut across the foot of the hill and meet you at the road," Erich added.

I left at a very brisk walk. The oncoming thunderhead, the eerie whisper of far-away wind, the odd yellow light that took the place of healthy daylight on the rocks and sagebrush . . . these quickly turned my brisk walk into a brisker run. When lightning hit the hill beside me, I began to run real fast.

Even then, habits are hard to break. I passed a bone, sticking up from the hard ground, a sizable bone, part of an ilium. I grabbed it up and went on but then noticed the edges of neural spines and sacral vertebrae sticking up out of the matrix. As I looked about hurriedly for loose rocks to mark the spot, my hair stood up a bit as lightning struck the far side of the hilltop; the place I stood might well be the next target. Somehow I got a skimpy rock pillar set up and then left under some pressure.

I reached the Buick just ahead of the rain. The trail where I had parked was very narrow; I thought I never would get the car turned about and headed for home. A few big drops stirred the thin dust on the windshield and splatted loudly on the rocks. This inspired me to get going; abruptly I was out of the canyon and headed south, away from the storm for the moment. The Buick's wheels bit better once I was out of the sand. Still ahead of the storm, I reached Erich waiting on the open road. Looking back, Erich saw the hill we had left washed out of sight by cloudburst. As we raced the rain to Rock Springs, we were astonished by the amount of water already pouring under the bridge where it usually crossed a dry river bed. Then we recalled the morning storm that hadn't touched us; the water it had dumped on the upper watershed of Baxter Basin was just reaching us now. As we paused to watch, the water continued to rise. Then the rain began to catch up to us.

"Let's cut out for the Howard Cafe!" Erich urged. "We'll be better off sitting it out inside."

It was only four o'clock, but the proprietor was already turning on the lights. "Looks like you fellows struck it lucky!" the waitress said as she set coffee before us. "But where's the rest of the boys? And where's Dr. Brown?"

Erich and I swapped open-mouthed glances. We had forgotten Brown! Gil and Don and Ted Lewis, working near the airport, were probably holed up safely at the airport or in camp. But

Brown? He and Bob Chaffee, far up north, might be in different circumstances. Even if they had tried to get away before the rain started, it was slow going down Eden Valley roads before they'd hit the paved highway at Winton.

"Good Lord," Erich groaned, "I'll bet they're still up Long Canyon! And if they are . . ."

We looked out the window, across the wide street, the Union Pacific tracks, the buildings across the street. First the buildings faded from sight, behind the rain curtain, then the tracks, then the street. Suddenly our side of the street—what we could see—became a river. The pounding rain and the racing water sounded like a cascade. The flood washed the curb, crossed the sidewalk, then paused at the edge of entering the restaurant. As the rain slacked off, we could see things again. Like the Buick, stolidly sitting there, but breasting the waves as if it were a boat.

"I'll bet they're getting it at Point of Rocks again," Erich said. "Along with that other storm, what this will do to Bitter Creek! Let's drive out past the coal company's offices and have a look at what's gone on." The rain had slacked off to a mere sprinkle.

At the curve of Bitter Creek quite a sight met our eyes. Men were frantically building a sandbag dike against the flood, only they weren't sticking strictly to sand. They were cramming bags with anything that would go into a bag, including horse feed! A nearby native remarked, "I've been forty years in Rock Springs, and I never see a cloudburst like this!"

"Wouldn't surprise me," another remarked, "if some of the houses below town got washed away."

"They say out in Eden Valley that Kilpecker Creek looks like the Mississippi—clean across the valley."

"And we've got to cross it to get back to camp," added Erich. "Guess we'll put up at the hotel tonight."

We turned around, crossed town, crossed the bridge over Bitter Creek, which was now the width of the valley. The road to Winton skirted the shore of this sullen flood, so we explored it toward the north. Long before we reached Pinedale Fork we were halted by inundated highway . . . as well as by the cars lined up ahead of us. "No use to try and get through tonight," a man called out. "Not even if the creek goes down; the whole roadbed's washed out this side of the bridge."

Back at the Park Hotel we learned the Union Pacific was out at Thayer Junction, near Point of Rocks. Two hours late, the evening train came backing into the station; the bridge at Horse Thief Canyon had gone out.

We started out for camp again in the morning. We knew we couldn't make it but wanted to show a proper attitude. Sure enough, the road was washed away at the bridge abutment, but the bridge still stood and bulldozers were shoring up the missing section. In an hour, we were able to get across and make our way back to camp. We met Don coming out of the cook tent.

"Look at the bad pennies turning up!" he greeted us.

"Where's Brown?" Erich asked.

"Brown? You expect to find him here? We spent half the night trying to get a rescue party across the valley. If anybody reaches that truck in Long Canyon in a week, I'll be surprised."

"Probably Brown and Bob have walked out to Winton by now," Gil added. "But where have you fellows been hiding out?"

We told our story and laid plans for the day. Don and Erich would drive up to Winton, looking for our missing persons. We would check out the plastered skull and fend off the press. We started off afoot for the airport but shortly holed up in the hangar when the rains came on again. Ted and Gil decided to go back to camp, rain or no rain. I waited it out, and about three o'clock started off toward White Mountain. Cutting in on my rambling reverie was the blast of a horn . . . the Buick, with Erich and Don aboard. They had found no trace of Brown or Bob Chaffee anywhere along the Winton road. They told me reporters had already picked up the disappearance. I could see it now: "Barnum Brown, Famed Dinosaur Hunter, Lost in Cloudburst."

Disconsolately, we rode back together to the rain-drenched camp. We parked and waded through the yellow mud, lifted the soggy tent flap. Maw Denniger was busy washing dishes.

"Well, they're back," she said. "Just et. Dr. Brown's gone to bed."

Bob appeared, smiling wanly and yawning widely. He and Brown had hitch-hiked in from Winton, somehow missing the searchers along the way. We got the story at breakfast next morning, from Brown. The rain had caught them plastering the big *Trachodon*.

"About nightfall the heavy rain stopped," Barnum told us. "It was almost freezing cold, and we had no coats. We made jackets out of gunny sacks and set out for Winton, hoping to wade the swollen stream, but found we didn't dare try. So we returned to the shelter of the truck. We didn't reach it until after midnight, and then we hurried around for wood to start a fire. We found it—the wood we had brought along for reinforcing plaster—but we had used all our matches lighting up our smokes."

"That's where Bob earned a medal in my book," Barnum went on. "He showed the touch of genius more scientists should have. He started the motor, jumped a spark from a plug to light a gasoline-soaked cigarette paper, and brought us the gift of fire. We built a fire, then, inside the truck, in plaster pans full of rock, and took turns tending it. That's why we're both only half sleepy."

21

FOR THREE DAYS Bob Chaffee had been chipping away in hard shale in a small quarry, high up in the sharp face of the cliff under the Point of Rocks beacon, following bone into the matrix, chiselling a shelf on which to stand. The specimen Erich had located on his first day was leading into a fine skeleton. Brown was especially interested, as it promised to represent still another type of dinosaur in the Mesaverde. There was but little bone visible, but he had reason to think it might be one of the ankylosaurs—the so-called armored dinosaurs, great, squat, four-legged creatures resembling giant horned toads. So far Bob had not come to any of the dermal armor that would make identification positive. At least, a good specimen of some sort was here, tightly encased in flint-like rock.

On the fourth day Erich and I stopped by to check the progress of the work. Bob was wielding hammer and chisel, splitting off the bedding plane above the hard layer. Fortunately the layer was fractured, meaning we could take up the bone-bearing mass in manageable blocks.

"How do you plan to get the blocks down to the road, Bob?" Erich asked.

Bob chipped away, thought awhile. "I plan on using a form of gravity, controlled gravity. Maybe with rope, maybe block and tackle."

Erich turned to me. "Personally, I'm glad we have an easier job."

Prospecting for new material has its drawbacks, but they don't include engineering problems. We seesawed in westerly direction, probing here, prodding there. The only find of the day: Erich came upon a weathered horn core lying free on a rocky ledge. Digging into the shale above, we came upon another ceratopsian skull. When we saw Brown at supper, he laid plans for removing it within the week. Don looked at me, his eyes twinkling.

"One day," he said, "somebody's going to find a lot of headless ceratopsians in the Mesaverde, and pick up an easy doctorate by making a new genus out of 'em."

Brown and Gil and Lewis lent a hand when it came time to lower the blocks of the armored dinosaur. For the sake of the reassembly, I first shot pictures of the specimen, with numbers painted on the blocks. Then, one by one, they were lowered two hundred feet down the exposure of Ericson sandstone in a washtub at the end of a long rope. The specimen was unique in that it took very little plastering; only here and there where bone protruded from matrix were patches needed.

While this job went on, Erich and Don and I took the Buick up the road toward Gunn, on the inner side of the Baxter Uplift. Prospects here had never seemed promising from a distance, but Brown had wanted them looked over, and this seemed a good time.

Gunn was a ghost town. Once a thriving community like Winton and Dines, it had run out of coal, and the company had moved the machinery away. The tall tipple under the brown sandstone cliff was abandoned to the swallows and the winds, its loose old boards here and there clattering disconsolately now and then. Its rows of little company houses, drab and grey and weathered, were deserted, except where here and there tenants hung on or squatters moved in. The main street was a mini-badlands of weeds and washes and gullies created by the recent cloudburst. Nobody cared. Instead, a new street had been opened through the backyards.

The old Buick creaked and groaned in every joint as we took the first turn on the new main street, missed an outhouse by an inch, tilted the bottom board on a set of ramshackle steps, and jerked sharply to miss a big hole in a sidewalk gutter. We passed the school, empty-eyed and idle. The government had removed its label from the dead letter depository but had left a naked flagpole. Black window-holes stared at us blindly. Gunn was dead.

We parked at the abandoned tipple. The rocks to the west, like the town, looked bare of anything worthwhile, but we were here. We trudged upward, where children had once played house. Rows of little stones marked off the rooms, and broken bottles and bits of broken chinaware lay randomly about.

The surrounding exposures, naked and baking in the glaring hot sun, were as uninspiring as the dead town of Gunn. They promised nothing. After hours in the heat probing and digging about, we felt the exposures had kept the promise. We were accustomed to disappointment in the Mesaverde, but somehow amid all this abandoned desolation this routine search was especially disappointing. It was a relief to go back to the car.

Brown and the boys, finished with the skull at Point of Rocks, went to view Erich's latest ceratopsian skull. "Looks as if Joe Louis had punched him in the snout," Brown commented. We still thought it worth collecting, though much of one side was missing and the jaw was entirely gone. Some museum might find it good for repair parts. Lewis and I took up the find I had made while running from the cloudburst, pelvis and hind legs of another ceratopsian. If you get enough odds and ends like this, you can make a dinosaur, so we piled it in with the rest of our plastered pieces.

The worst part of what August heat and rains and intermittent dust storms did was to rub Maw Denniger raw. Came an afternoon when we came in for supper to be met by Mrs. Brown over the stove. "Maw went to town with me today," she told us, "and saw a bus headed for Montana. That's the end of Maw for this trip." Mrs. Brown lifted the lid from a savory stew and told us, "I guess I can cook for our gang. 'Til we find another real cook."

In fact, we could have done well on the whole expedition had we never had anyone but the Browns as cooks. On Monday, August ninth, we were routed out of bed by unusually loud clanging on the tire iron at the end of the cook shed. Brown stood over the stove absorbed in his cooking. He did, in fact, enjoy the work and often took over on Sunday mornings. It made me hungry to watch him; there was something about the way he flipped the eggs and forked the crisp bacon from the pan that made his work visually appealing.

The boys came filing in to the table, arguing good-naturedly about whose turn it was to ride in the back of the truck, known as the Cannonball Cloudburst Special since the big flood. There wasn't room for everybody on the front seat, so usually at least one or more of us got bounced around in the back.

"I don't see why it has always to be me in the dog cage every day," Don complained, "just because I take up more room. T'aint fair; it's not my fault my pituitary ran amok."

"What do you mean, you get stuck every day?" Bob asked. "If my memory serves, I rode in back yesterday."

"I think it's Erich's turn today," Lewis cut in.

"That brings up a subject I've been kicking around," Brown interrupted. "We're now going so far afield, we scouted the immediate area so well that we're putting in more than half our time riding to work or coming back to bed, leaving a few hours in the hottest part of the day to dig. Time is running out on us, and we've got to use it better."

He turned to Erich. "How about you taking a party and set up camp up around Chilton's Ranch? Take supplies for a week, and take a week. Take Gil and Bob . . . and big Don too, if you promise not to lose him . . . and see what you can do with the time you save."

He looked around. "Lewis, let's you and I and R.T. take the Buick and work some leftover spots we've missed to the south of here. Stuff that should be looked over, but there isn't enough work for all of us."

Shortly after breakfast, the two parties took off, the Buick piloted by Ted Lewis. We swung south of Rock Springs and turned east at the colliery. We passed under the Golden Wall, cut out into Baxter Basin, turned west again to climb up over the brown spread of the Golden Wall and the grey Ericson sandstone.

"Erosion has just about washed out the Mesaverde here," Lewis noted.

There were, indeed, few favorable exposures. But I noticed a low ridge some distance away on our left, a thin outcrop of sandstone, with traces of other sandstones beneath. I pointed it out to Barnum, who studied it carefully.

"I'd like to play my hunch on that," I said.

Lewis slowed the Buick. Brown still studied the outcrop.

"I suppose you might as well play your hunch," he said. "Go ahead, R.T. Lewis and I will circle west and meet you back at the car for lunch."

Ahead, in a sea of scattered sagebrush, the ridge rose like a solitary island in a bone-dry sea. The ledges that had attracted me looked even better close up. An ancient stream channel was represented in the upper ledge. I reached the apron of the hill and could almost smell dinosaurs. I picked up a small clam shell. A bit of fossil wood. A sizable bit of bone fallen from higher up. I clawed my way upward as fast as I could toward the lenticular sandstone. Just under the edge several more fragments of bone lay about the surface, their size indicating a very large animal. Part of a limb bone and a few ribs protruded from the greyish shale. I picked up and recognized the weathered end of a truly huge humerus, the piece that made contact with the scapula and coracoid, a rounded piece. I couldn't name the creature, but assuredly it was not a ceratopsian. Other parts of the broken shaft and head led into the bank a few feet below the sandstone.

I began stripping away the surface of the overlying matrix with my hand tools. The point of the awl scraped sharply against a new object; with the humerus and broad rib ends I now came upon radius and ulna. I dug and broomed away matrix until the sun was high in the sky. Clearly I couldn't finish before lunch. I picked up the rounded head of the huge humerus and headed for the Buick, three quarters of a mile away.

A hundred yards from the car I put a hand, with the bone end, behind my back, but Brown noticed me and grinned.

"Bet I can guess what you got in your hand," he said. When I handed over the rounded mass of fossil bone, he whistled long and low.

"Why, R.T.! This is the left humerus of an iguanodont, almost large enough to be our mystery dinosaur, our big fellow. Is there more?"

I described the outcrop.

"Eat your food," he ordered. "Don't stand on ceremony; go ahead and wolf it down. Then we'll all go over and help you discover things."

While I hurried lunch, Brown and Lewis gathered up their tools, filled a waterbag, and we were off together. On the ridge, Barnum took in the situation at a glance. "Now we have manpower," he suggested, "let's clear away a bit more overburden."

We shovelled off a few feet of loose brown shale. I let Barnum take the big humerus, while I went to work on the ribs. Lewis took over the radius and ulna. The matrix was soft; the bone, hard. Much better than we were accustomed to in the Mesaverde.

The hours flew by. A breeze sprang up to temper the midday heat, its cooling sweep uninterrupted by any other ridges. Brown, whistling while he worked, was happier than I'd seen him in weeks.

The humerus took on shape. The ribs clearly belonged to the same animal. And the radius and ulna. They were no longer bones to me; my imagination decked them out in flesh and blood and I could picture this great iguanodont, some thirty feet high, kangarooing his way down the hill, galumphing across the sagebrush flat, alternately using his tail as support and balance. This was just before Brown said, "I've run out of humerus."

Lewis, too, was running out of the bones he was working on. The matrix continued to grow harder and harder, and then ceased to be matrix and just turned to hard rock; it's only matrix when it has something in it. Desperately we cleared away more overburden, dug farther back into the bank, hoping against hope that more of our creature was hiding there. At last Brown laid down his tools with a sigh.

"Well, R.T.," he groaned, "this must be it."

A let-down feeling seemed to go with the Mesaverde somehow. We looked at what we had. It was good, just not good enough. The bones were fine, as far as they went. We shellacked them. We'd return tomorrow and plaster them. We'd take them home. After all, if you get enough pieces you can always make a dinosaur.

"I've got to see that humerus put together," Brown said. "It certainly looks more like the big fellow than anything we've seen."

Next morning we plastered the bones. To give them a proper time to set before taking them up, we decided to scout a bit more to the south. Not because it was encouraging, but because it was there. We passed through a canyon in the Ericson sandstone and a green little valley opened up before us, a valley green and lovely with alfalfa, surrounded entirely by grey cliff walls and well-watered. Nothing in the West is so overwhelming as a solid green carpet covering a smooth valley floor. There was a neat white ranch house with sheep and cattle grazing about for scenic effect and probably for use later.

"Say, this is Cappen Ranch. I know this Cappen; I was going to visit him before we went back anyhow. I'd say now is the hour. Cappen has a collection of Indian stuff that . . . well, he is a museum in his own right."

Cappen and his boys were turning freshly cut alfalfa in a field along the way. We stopped to meet him, and Brown asked about his artifacts.

"We'll soon be through here," he said. "If you'd like to go to the house, we'll be there in a bit."

As we cut across the stubble field, Brown called attention to the fact this valley had been well-used long before the white man arrived here. There were springs of flowing water, protection from the winds, good ground, wood for fires in the gnarled cedars under the Ericson walls, a great cave in the canyon wall to make an excellent rock shelter. All these things indicated an abundance of artifacts.

We idled along the way, so reached the ranch house not much ahead of Cappen and his boys. Mr. Cappen showed us into the living room, and we could see the truth of Brown's statement. The walls, the backs of chairs, the seats of sofas were draped and hung and covered with cardboard and plywood plaques, each adorned to the limit with arrowheads, spear points, flat stone knives, scrapers, and awls for punching holes in leather. All three of us were enchanted and covered the collection thoroughly. As he stood before the last plaque, Brown said, "I've never seen a finer collection . . . not in a museum."

Cappen smiled in modest pleasure at Brown's flattering but sincere remark. He dug out a cigar box from a table drawer. "These aren't good enough to mount," he said, "but since you're so interested . . ." The boys brought out similar cigar boxes, and Brown sat down and carefully fingered every last point. I knew what he was looking for.

"You've never picked up any Folsom points here?" he asked.

No, the Cappens said, they had not. The known history of the Folsom point was as intriguing as it was limited. Barnum Brown had himself been the first man to find and to give name to the Folsom point. On a day in the early 1900s he discovered a strange arrowhead or spearpoint lodged in the bones of a prehistoric bison he was excavating. Subsequently, from other skeletons in the same quarry more of these points of a hitherto unknown type were brought to light. One, deeply imbedded in a vertebra, proved it had contributed to the creature's demise. Obviously, Brown had made an important archaeological discovery.

Folsom points, either as arrowpoints or as spearheads, were almost mechanically perfect in design. Unlike ordinary points, triangle-shaped, they were modern in line as a streamlined train. The piercing edges, curving back from the point, were like the edges of a double-edged knife. A distinctive feature was the long, shallow grooves on either side, providing a much better means to attach the wooden shaft of the arrow. The age of the bison skeleton indicated the points to have been fashioned by a race that antedated by thousands of years the people the first Europeans knew as Indians. Since the opening of that historic quarry in Folsom, New Mexico, collectors continue to turn up more of these significant missiles from widely scattered parts of the country. But their occurrence is rare, and Brown was always on the lookout for new specimens.

On our way out of the valley we stopped at the cave under the canyon wall where the Cappens had found some of their collection. The overhanging roof, black with smoke of long-gone fires, spoke of ancient days. We kicked about the debris on the floor but of course expected to find nothing. The Cappens were competent artifact hunters.

In late afternoon we got back to the hill where our plastered iguanodont bones were drying. Near the little quarry Brown picked up a cylindrical bit of sandstone, an inch or better in diameter, perhaps six inches long. He laid it in my hand. I passed it on to Lewis.

"That," said Brown, "is from inside a limb bone of a carnivorous dinosaur." The discovery added one more type of dinosaur to the Mesaverde record.

By week's end we had fairly well cleaned house south of Rock Springs. No new material had been found. Saturday night Erich and the boys pulled in from the north, reporting a completely articulated skeleton from a place near Chilton's Ranch. A *Trachodon*, he said, not very large but with bones in excellent condition. "We've started taking it up," Erich said, "but we want you to come look at it before we go any further."

"We'll drive up tomorrow," Brown replied.

I went to the quarry with Erich, Barnum and

Mrs. Brown, while the rest took Sunday off from bone-hunting. They gave themselves a busman's holiday, going in search of Indian artifacts to the south instead. Erich's specimen was everything he had said it was, occurring in soft sandstone, and its packing and removal were decided upon. It would stand, mounted, about twelve feet high.

Back in camp the artifact-hunting party had returned by suppertime. Lewis was elated over a find of his, a flint instrument, perfect in every detail, patinated from long lying in the sun. I recognized it instantly. So did Brown. His face shone with delight as Lewis dropped it in his hand. It was a rare and beautiful Folsom point!

22

BROWN, as early as June, had talked about taking time to visit Mesaverde exposures near Medicine Bow, Wyoming. Partly a nostalgia trip, perhaps; over forty years before he had captured his first dinosaur near there, from the Morrison Jurassic. This big fellow, a *Diplodocus* dug out in 1897, was the first dinosaur of any description to stand in the American Museum of Natural History.

"We can do this while the boys wrap up Erich's *Trachodon*," he suggested. "We can even have a look at the Cloverly, near my old dig at Como Bluff . . . and we can drop by Bone Cabin Quarry, too." Bone Cabin Quarry was famed for another great dinosaur, the huge brontosaur in the American Museum.

On August 16 we began the two-hundred-mile drive to Medicine Bow, arriving midafternoon. I slowed up before the Virginian Hotel, expecting we'd stop and register. "No," Barnum said, "let's drive right on to Como Bluff, like getting back to a boyhood home."

The road rounded a turn and approached a long bare ridge rising from the desert. Bare desert. Desert glowing with the familiar pastels. Light greys and dark greys. A band of soft purple, bluish purple, running the length of the bluff. A trace of maroon above the purple. As readable as a printed page; the variegated shales and clays were a page out of Jurassic Morrison time, a time when it was quite fashionable for dinosaurs to lie down and fossilize. Brown leaned forward in the seat, eyes on the approaching bluff. A curve indicated we should soon pass along the flats at the south end. As we neared a sagebrush-covered embankment he said, "There! That's the old roadbed of the Union Pacific; they've moved the track since my days here."

At a dirt road swinging off into the desert, in the same general direction as the embankment, Barnum said, "Let's take this turn."

I wheeled the Buick over.

"That's the quarry, yonder!" A gash, tiny in the distance, was visible in the side of Como Bluff. Barnum was tickled as a boy on the Fourth of July. "Every day the engineer on the morning train threw off my mail as he went by. He'd always toot, to let me know he was coming."

We came to a turn leading directly into the bluff. Brown chuckled. "I had a horse run away from me once, right here. With a loadful of bones."

A turn-off led to the quarry. As I prepared to swing into it, Barnum suggested we first drive over to the ridge. We came to an outcrop of brown sandstone. That would be the Dakota, hard, more erosion-resistant than the shales, forming the crest of the ridge. The Buick rolled over the top and wheezed easier, headed downgrade. Pink bands of clay appeared along the falling dip slope. "That's the Cloverly, what there is of it," Brown commented. The beds were the equivalent of the gaudy Cloverly at Howe Ranch Quarry, but they lacked vividness and variety here. Brown watched the wayside carefully as we coasted by. The trail approached a filling station, and we coasted in; the Buick was getting thirsty.

An attendant came out to fill our tank, walking by a huge sauropod femur set in a concrete panel, dark grey, nearly black, like those of Howe Quarry. A few yards away stood a small building, concrete, but into its walls had been set more of the same dark bone. It looked as if a brontosaur had gone all to pieces there and the builder, lacking better building material, had stuccoed the walls with the resulting rubble . . . ends of vertebrae, shafts of limb bones, pieces of ribs, broken ilia and scapulae.

I followed Brown to the odd little bone building, a museum owned by a Mr. Boylan, proprietor of the station. Barnum introduced himself to the genial, middle-aged gentleman.

"Barnum Brown!" he exclaimed. "Why, I feel like I've known you for years!"

"Then you've heard about my old quarry up on the bluff?"

"Heard of it? I even put a sign up there myself years ago, to mark the spot. Have you seen it?"

"We've just driven by," Barnum said. "We'll go back and look it over in a few minutes."

Boylan wanted to accompany us, and after he had shown off the various fossils he had on display, he piled into the Buick and we were off. We were soon following the little trail in to Brown's old quarry. Not far from a fallen block of Dakota sandstone stood a stout post bearing Boylan's sign. I found a handy place to park and went to read the sign, "Barnum Brown took a dinosaur out of here in 1897."

What had seemed only a slight gash in the hillside from a distance turned out to be a fairly impressive hole close up. Brown went over to inspect a pile of timbers and boards. In this dry climate, wood weathers well. "These are from the tunnel I had to dig to get the tail out," he said. He walked over to the outer edge of the quarry. "A big femur weathered out here," Barnum said, and added, "Right there was the dorsal region. Quite a pile. Quite a while ago."

I set up my old five-by-seven camera on a tripod and shot pictures of Barnum as he reminisced.

"It was in this quarry that I first introduced plaster of Paris in taking out fossils. Up until then, no one had worked out a technique for taking up dinosaur bones that didn't involve a good bit of breakage. I encased the bones in plaster and reinforced the jackets with leather thongs. Wood came later. The bones got back to the museum in A-1 condition."

When Brown suggested going to see Bone Cabin Quarry, Boylan broke in, "I've got a trout farm up there; like to see it? I'd be glad to go along and show it off to you."

At Boylan's farm we found a big spring pouring out from Triassic sandstone to form a series of ponds, ponds holding thousands of trout of all sizes darting about. Brown was intrigued.

"If I ever retire," he enthused, "this is a set-up I'd go for!" This was something to chuckle over; Barnum would be planning expeditions and digging bones on the other side of the Great Divide. I hoped I could go along even then.

We watched the trout flick their happy little tails in the clear water until it grew dark. No time now or no point to visit Bone Cabin Quarry. We went back to Medicine Bow and checked in at the hotel. The Virginian was named in honor of Owen Wister's famous book, set in this part of Wyoming. In the lobby was a portrait of Gary Cooper, who had played the lead in the movie shot in part in and around the town.

The exposures we had come to see, the local Mesaverde, lay about nine miles south of Medicine Bow. We were on the spot at eight o'clock.

It took until four o'clock to move from the exhilaration of treading virgin territory to sodden acceptance of the Mesaverde as a formation always capable of living down to its expectations. Walking in from the Cloverly beds near Como Bluff, empty-handed but for scraps, thoroughly discouraged, Barnum said, "Let's pick up Boylan and go see Bone Cabin Quarry. This is not a good day."

At the end of a long stretch of straight road, Boylan turned me in at the next left fork. The terrain dipped slightly. The road was running on Morrison, but the exposures, partly overgrown, lacked the vividness of the colors at Como Bluff. We came upon a large pit, overgrown with weeds and full of rubble. Bone fragments, like those in the walls of Boylan's little museum, littered the ground. I pulled at a mound of clay dug from the pit.

"Granger and Wortman worked here three seasons," Brown told us. "When they finished, they were just getting down to good bones."

The excavation covered an area about the size of Howe Quarry. It looked small for three seasons. "Three seasons?" I asked.

Barnum, climbing out of the car to look the excavation over at close range, chuckled. "Three. Walter was hell for duck-hunting. All his spare time, he was down at Como Lake with a shotgun."

I had to smile, thinking of Walter Granger, one of the most easy-going men I ever met, digging dinosaurs only until the ducks came by. But our first *Brontosaurus*, perhaps the most spectacular exhibit in the museum, came from here largely through Walter Granger's efforts, ducks or no ducks.

Down the hill below the quarry, in a ravine cut by recent erosion, Brown showed me what he had meant by just getting down to good bones. The Morrison shales, cross-sectioned, carried a solid layer of black bones for several feet. The whole hillside underground was filled with a litter of broken dinosaur bones. I was sorry we didn't have cause to open another quarry, but so much was already known about Morrison dinosaurs collected from this region, there was no point in such a project.

On Friday, August 20, Brown's old friends from Montana, George Shea and his wife, paid us a visit at the Rock Springs camp. This was just after Brown and I returned from Medicine Bow, and while Erich and his boys were still up north wrapping up the *Trachodon*. Brown thought it a good day to visit the outcrop of Cretaceous plants and shells Don had discovered earlier in the season. It was up in the Long Canyon area, but considerably higher stratigraphically than the bone-bearing layer. We decided to take both the Buick and Shea's car.

Driving to the base of the hill where the outcrops were located, Brown led the way on foot to the crest of the ridge. Here a band of chalky white matrix, soft but slatey, was interbedded with brownish-grey shales. Here and there lumps had been pried out from the weathering edge, probably where Don had taken his first plant specimens. Farther up was the shell layer I had heard of, a seam of sandy matrix varying from a few inches to a few feet, shot through with thousands upon thousands of fossil snails. They were so neat and clean, most of them so undamaged, that we set about collecting them. While we were disposing of our picnic lunch, someone observed that Don Guadagni had also found a fossil oyster bed just a bit farther down. "When we finish with the snails and the plants," Barnum suggested, "we should try our hand at oysters."

The plant fossils were a pure delight—a chalky shale with beautiful bedding planes and the plant prints beautifully highlighted as thin, shiny carbon layers, seams of coal running usually less than half a millimeter thick. We don't find eye-catching plant fossils such as these everywhere nor everyday. Every block we split gave us choices. Grasses, sedges, palms. It was hard not to go berserk on simply piling up plant fossils and to strive for variety, to discard good material simply because we already had too much of it.

When we moved on to the oyster layer things were almost as fine. Many of these ancient oyster shells were as pure and clean as if they were left over from a dinner of oysters on the half shell rather than records of marine lives dead and gone a hundred million years. So fine that we soon got persnickety and looked only for fossils with both halves present.

Mrs. Brown was working alongside me. "Wouldn't it be a good publicity thing, R.T.," she asked, "if when we go back to Rock Springs tonight, maybe we could have oysters on the half-shell? And maybe take along a few of these shells for comparison; how old did you say these are? Maybe I could make something of it, like a short article or something. You know, comparing these with fresh oysters one way and another."

"Sounds feasible . . . interesting . . . if they have fresh oysters in Rock Springs now. Remember, Rock Springs isn't nearly as near the sea now."

"Well, if the oysters couldn't be flown in, but now . . ."

"If you try it," I answered, "I'd like to suggest a title."

"What's your title, R.T.?"

"'Oysters Now and Then,'" I said. "'Oysters Long Ago and On the Half-Shell.' 'Oysters Once and Again.' You can try it out different ways, for euphony."

23

IT RAINED ALL DAY on September 8, which was the sour cherry atop a daub of sour cream. Mud Springs, where we had been trying to work, "was named by a local with an aptitude for accurate nomenclature," Brown remarked. "We can stay here mired down in this clay country for days. Or we can use the paved road and run away from all this. We've got to have a look at the Bridger anyhow, before we go home. Let's look at it now."

The idea wore well. There might even be better weather a bit west. Baxter Basin, after all, was giving us a case of outdoor cabin fever, and the Bridger Formation would be a change of diet fossil-wise. Not that we were bored with dinosaurs; it was just that after a diet of something heavy one is intrigued by a bit of parsley or perhaps a carrot stick. Bridger was above the Mesozoic, above the Age of Dinosaurs. Up in the time of the mammals. It would be nice, for a change, to find bones we could lift, that is, if we found bones. Like the early horse, the oreodont, even one of the early rhinos. The latter were big enough, but compared with dinosaurs they would look manageable.

Taking both truck and Buick, we left Rock Springs behind us. Ahead to our right White Mountain slowly rose out of the plain. A greenish grey cliff, with outcrops of reddish grey sandstone, rising above the Mesaverde, took us into Tertiary time. We passed the pinkish Wasatch, sort of a foothill to White Mountain. The road followed the cliff into Green River, named for the turbulent stream we crossed. Just beyond town the highway climbed the cliff across the face of the Green River shale and led us on past rounded hills, the first available vestiges of the Bridger. At first they were vestiges only, but gradually the greenish grey beds grew more abundant, the colors more vivid. A

long ridge on our left was well-exposed and looked to us like good hunting ground, but Brown urged us on.

"Granger collected here a couple of seasons," he told us. "Other collectors have come and gone; this has been pretty well worked over." Finally he spotted a site he felt looked worth a look. "We'll try it here until lunch time; if we don't hit something by then, we'll shove along."

It was nearly eleven o'clock then. We each picked out a section we could cover in an hour and went our separate ways. My chosen route took in a few low mounds around a small butte, bare and weathered, bare haystacks much like some of the Cloverly around Howe Ranch. The surfaces were hard and crusty, covered with loose pebbles that made walking tricky. It was, however, possible to look over several yards on both sides at a fair pace. I could just see myself coming upon a slender bone from the little *Eohippus*, the dawn horse. It was a good time and place for Lady Luck to deal us a fresh hand. The *Eohippus*, such a dainty little thing with his full set of toes and about the size of a fox, was so much more endearing than the great dinosaurs. And here they had gambolled about in the sun when these green beds were water-borne sands.

The first bone fragment I came upon was flat and roughly rectangular, and shortly I came upon several like pieces that seemed to fit together. Looking for a lovely little horse, a dainty mammal, I had come upon the leftover covering of a prosaic turtle. The turtle is one of God's best time-travellers; he was here before the dinosaurs came on the scene; he survived their passing. But somehow, because they were so common all through their time, turtle fossils are not much more exciting than turtle lives must be. I looked about the top of the stone haystack and found half a dozen piles of long-unused turtle carapaces, fellows that for whatever cause had all died here together.

Brown came up to me. "Any luck?" I asked.

"Turtle fragments." He smiled wearily.

When we gathered back at the car for lunch, Bob and Gil and Dan had shared our own luck. Lewis brought in the broken end of a small femur and a few assorted limb bones of an animal that couldn't have come from any creature larger than a dog.

Barnum examined them. "Possible oreodont," he hazarded a guess. "Hard to say, from the meagerness of the lot."

Little fellows, about the size of the early horse, oreodonts had in places almost infested Bridger time. From the plenitude of oreodont fossils, they travelled in packs or herds over the ancient plains, apparently much more of a success than *Eohippus*. What strange failing or event wiped them out, we will probably never know.

"I'd like to find a good complete specimen," Lewis remarked. "They tell me they're a lot easier to take up than dinosaurs, on the average, that is."

Next day was pretty much a repetition. Gil found a fossil snake, the only thing worth collecting.

"We might as well throw in our cards here," Brown finally decided. "We might go back and look over that Hiawatha region, maybe using that old sheep shed for quarters. This country has been too well gone over and too recently. There hasn't been time for new material to weather out since the fossil-hunters were last here."

This made sense. Our crew was steadily shrinking, due to the beginning of college fall terms. Mrs. Brown had left us again for California. We were now down to Barnum and Ted Lewis and me, and Lewis was shortly to head back to the university.

Roads were getting bad with the onslaught of fall, and it was night before we drew up beside the old sheep shed outside of Hiawatha. The wind was rising and the thermometer dropping. Lonely as the spot was, we were glad to get into the old shed. Inside, the lantern lit up the neat bareness of a large kitchen, a big cooking range in the corner, two long tables for feeding sheep-shearing crews. While we dragged in cots and blankets and our mess box, Barnum got supper going on our camp stove.

It was a spooky spot. The rising and unrelenting wind made the rafters of the old shed creak and groan. A loose piece of corrugated roofing alternately snapped at the roof and squealed in agony. Darkness outside was, of course, absolute, and a gasoline lantern somehow creates a special quality of darkness just beyond its range.

Lewis, less accustomed to western ways, said, "I don't like this. The place here is empty and open, of course. But there must be a caretaker about somewhere. What's he going to say when he comes around and finds us making free with his property?"

Almost on cue, the lights of a car swung in from the road and caught the shed. Barnum was unmoved. He poured a spot of cold water into the coffee to settle the grounds. The car stopped, but its motor was left running. The old gate, which had creaked mournfully heretofore, now sounded authoritatively.

The kitchen door burst open and the wind blew in a stocky little man with a sandy mustache, wearing a mackinaw buttoned tightly about his neck. He carefully caught the wind-whipped door and closed it behind him.

"Pretty windy night out, isn't it?" Brown said briskly. He set down a pan of potatoes and held out a hand.

"The name is Brown," he added. "A fine kitchen you have here."

The newcomer blinked in the glare of our gasoline lantern. He looked around at each of us, at our meal in preparation, at Brown's hand. Then he took it.

"Well," he said, "I'm glad you like it. But your lights gave me a start."

Our caller, who introduced himself as Sandy MacGregor, made us welcome when Brown explained what we were doing in the area. He told us where to find water and told us to use the shed at Cow Creek Ranch as long as we needed. Brown invited him to supper, but he declined and hurried off.

Next morning we drove in to Hiawatha to look up Jack Deamer, with whom Brown had had correspondence about fossils. Deamer was enthusiastic about the places where he had found the bones he had told Brown about. He took us out to look over assorted bones he had found in the region. They were indeed assorted but ran mainly to titanothere, a creature that might be termed a distant cousin of the rhino. Along with many creatures of their time, they were tremendously successful but failed to survive. Deamer had come upon a fair skull and gave us directions to the place where it had been found.

His excellent directions led us out across flat terrain to a break in it that became a canyon cut into the underlying formations, formations rich and bright with Bridger colors. This widened into a veritable sea of badlands, miles of erosion-sculptured rocks, moss green streaked with greys and browns. On foot we climbed down into a wide amphitheatre and into a strange world. The canyon walls resembled walls of a tumbledown temple, a pillar missing here, a cornice there. We felt more like archeologists picking our way through ancient Greece. We prospected the area all afternoon, and came back the next day. Mammal bones were fairly common but unfairly fragmented. Turtles were abundant; we ceased, in fact, to consider them collectible fossils and took them as part of the landscape. One turtle, however, was so well integrated that he attracted Barnum's attention as a collectible, and this led to discovery of a complete titanothere, good enough to be a mate for the skull on Jack Deamer's porch.

Next morning I noticed Barnum limping as he went about the big shed preparing breakfast.

"Did you hurt your foot?" I asked.

He made a wry face as he stepped again on the bad foot. "I wrenched it yesterday evening coming up out of Skull Creek. Didn't pay any mind to it, but she's a lot worse this morning."

After breakfast the wrenched ankle seemed no better. He gathered up his tools to put them in the car, but at the last moment took them back out. "R.T.," he said, "I hate to do it, but I'll just simply have to stay in camp today."

Lewis and I drove out to a new location along Skull Creek, where the beds looked low stratigraphically but struck us as favorable for bone. We clambered down to a point below some large channel sandstones, rocks eroded into a weird forest of stone toadstools. Crossing a few dry washes, we climbed a long ridge of green sandstone, where we came upon a tremendous petrified log. It protruded nearly its full length from a sandstone ledge, nearly four feet through at the base, about twenty feet long. Under it, supporting it, was a clay bench.

"I saw a log almost as unusual as this the other day," Lewis remarked. "And I don't think it's very far from here."

We set out in what he thought was the right direction, around an intervening butte. In the weathered face of the exposure were lodged three tremendous limb bones, bones of one of the largest uintatheres of Bridger times. The creature would have rivalled the largest living rhino; had I not spent some years digging up dinosaurs, I'd have been greatly impressed.

"What a great nuisance!" Lewis exclaimed. "Now we'll have to go for plaster when we go back for lunch."

This was to be Ted Lewis's last day in the field, so we trudged on despite this setback. Ahead was an odd little hillock, perhaps fifteen feet high, maybe fifty feet across, nicely rounded, a curious formation. When we clambered to the top, a curious sight met our eyes: it was completely capped with fossil shells of big and little turtles, some of them up to three feet across, all obviously having died together much as the dinosaurs at Howe Quarry. Though most of them had been crushed, the carapaces had served as a caprock for the odd little butte.

In late afternoon we trudged up through the stone toadstool forest laden down with our uintathere bones all nicely plastered. A fine finale for Ted's summer, but more than a bit fatiguing. We got back to the sheep shed just before dark. There in the yard Brown was tugging way, dragging plastered bones from the Buick, among them a fine titanothere jaw.

"Where did those come from?" Lewis asked. "I thought you were a cripple."

Brown grinned sheepishly. "I got them with Sandy MacGregor." Sandy had talked him into going for a look at a bit of the Bridger he was acquainted with, and Brown, despite his bad ankle, had ended with as good a day as we had had.

WE MADE JACK RYAN a member of the party when Maw left the cook stove to go home to Montana, when Mrs. Brown left us for relatives in California, when the oncoming fall term took the rest of our boys back to the college circuit. Jack wasn't as lovable as Maw, but he could cook. He could also wield a hammer, drive a truck, lend a hand in a dozen different ways.

In the tag end of September, a time when summer is clearly over in the West, Jack took the truck and I mounted the company Buick to set out for Colorado and, ultimately, to catch up with the mystery dinosaur footprints at Cedaredge. Brown had gone on ahead by a few days on business in Salt Lake City, and we would get the show all together again in Grand Junction. Our season's boxed and plastered bones were stored in a Rock Springs warehouse. On the previous trip to Cedaredge with Brown, we had gone by way of eastern Colorado. This time, for reasons of nostalgia, I chose to backtrack along some of the route followed on my first trip through this country by motorcycle. Then I had been just a biker gawking at the scenery. Now I had been exposed to enough geology so that the hills had something to say to me. We dropped with the road down below a familiar grey scarp in Willow Creek Canyon. Now the faces of the side draws were Green River shales, now we were crossing the Tertiary, now we were watching for the Mesaverde beds to appear. We came upon them in the lower end of Price River Canyon at a place called Castlegate. They were no longer just rock; now the Mesaverde was a familiar old friend, an often disappointing old friend, but one we were glad to see again just because we were no longer strangers. Ryan was in the lead, and the back of the truck's red body swung around one curve as the Buick swung into another.

A few miles beyond, at Price, we spent the night. Next morning we swung east on U.S. 50 toward Green River, Utah. Here sixty-five miles below Price, the road turned a corner under one of the massive cliffs north of town.

These were the Book Cliffs. An observant child could have recognized in the laminated walls a likeness to the same sort of coal formations rising around Baxter Basin now two hundred miles northeast. The grey Mancos shale below was as easily recognized. The town of Price rested upon it, surrounded by irrigated alfalfa fields. Green River stood upon it too, and miles of empty desert stretching south were Mancos shale as well. Older Mesozoic formations were visible in the far distance in the San Rafael Swell, but except for a glimpse of the Jurassic Morrison, most of them were too far away to identify.

The route passed east through Thompson and Cisno. The march of the Mesaverde, now on our north, was continuous. We stopped for a few photographs, and Ryan observed, "The Mesaverde is going to beat us in to Cedaredge yet."

Brown was in La Court Hotel in Grand Junction, waiting for a long distance phone call. "You boys might as well go right on up to the mine and get established," he told us. "See you in the morning."

The clay road leading on up to the States Mine was bright with the afternoon sun, brilliant green with the quaking aspen under the cliff of Rollins Sandstone, still untouched by fall. Alfalfa fields were a brilliant patchwork across Green Valley. The stark grey tipple outlined against Grand Mesa contrasted sharply. Blue smoke still drifted lazily from the slag pile, tainting the air mildly, probably slightly poisonous but not unpleasant. Mr. States was in his office.

"I'm certainly glad to see you boys," he greeted us. "Where's Dr. Brown?"

It was agreed Ryan and I might camp in the bunkhouse. Jack soon had supper going on the gasoline stove. He lacked Maw's motherly ways, but his cooking was good enough. And we were both hungry.

"How long do you reckon this job'll take?" he asked. "States seemed concerned about that forty-inch sandstone ceiling, didn't he? Cutting down a block eighteen feet long and six feet wide through that stuff sounds like all winter, huh?"

Brown showed up early in the morning, and as soon as States could leave the office we started for the mine entrance. The temperature inside the entry leading to the long maze of tunnels was the same as it had been in the hot days of summer. The pin-point spark materialized out of darkness to turn into a miner's lamp, the glint of a lantern against sparkling spots in the coal seams, the odd sense of walking under a tremendous span of time, all these were no less exciting because we had done it all before. To Jack Ryan, this was an old story lacking in glamor; he had worked as a miner.

When we arrived at the room with the big footprints, we found a great pile of rock on the floor. Miners had already made a hole in the ceiling just beyond the point where Brown was planning to cut the block. States climbed the pile and peered up into the gaping hole. Brown and I crowded after him. The section cut out of the ceiling showed that what we had supposed or hoped would be a mere eight to ten inches of sandstone had turned out here to be forty inches thick. Our lights caught the edge of the cavity mined in the coal seam above. It was three feet thick. "That coal is going to be our salvation," Brown told us. "but we'll have to reduce the thickness of the ceiling from above before we try lowering the slab. How many miners can you spare us for this job?"

States said he thought he could round up seven or eight men. As we stepped down from the pile of coal and sandstone, a lean-faced miner, long and lanky, appeared in a nearby entry. Bill Fogg, the mine foreman. States introduced him. "I've been wanting to meet the man who'd dream up a job like this and then tackle it," he told Brown.

The two discussed mining the coal above the ceiling, propping the room, the manner of lowering the heavy slab. Brown told Fogg to go ahead at once, working the men in three shifts. Then we turned to look over the great tracks again. Long

and futile search for the tracks' maker only made them in some ways all the more to be desired.

"R.T., I've got to go back to Grand Junction for a short spell. It'll be up to you to stick around and oversee things. I'll be back in a day or two to see how you're getting along."

Two younger brothers of Bill Fogg and a husky fellow named Roy Aldrich went to work at once with the cramping job of mining the coal above the ceiling. Roy and the mine foreman began to set heavy cedar posts around the white chalk line that marked out the edge of the slab. The heavy props were set in about four feet apart, for the weight of the forty-inch-thick ceiling would be something to reckon with, once the sandstone was weakened by a long cut. Fogg next set up a coal augur between the two props. This is an augur-like bit about two inches in diameter and six to eight feet long, used to drill holes for blasting in the face of a coal seam. It was turned by hand with gears and a crank.

"We'll cut holes along the chalk line so close they'll almost overlap," Fogg explained. "But it's going to be a pile of work to go entirely around this block."

I tried my hand at this crank and found it to be an excellent memory aid; I immediately recalled Brown wanted the whole operation covered with pictures, so I went for my camera.

The work went on around the clock. As the line of holes grew, Fogg set up timbers to give support to the ponderous block when it should be freed. Brown came out on the third day and expressed satisfaction with the way the job was coming along.

My days were full. When not attending to other details, I explored the maze of tunnels, finding here and there prints of fossil plants to add to our collection. While I was pecking away at some small fern fossils near a place where miners were at work, Fogg came upon me.

"There was a time I could have showed you some plants that'd make those look like two cents," he remarked.

I paused with uplifted hammer. "Where's that?"

Bill chuckled. "I only said there was a time. We used to call it the picture room."

"The picture room?"

"Yes. It was in the old Red Mountain Mine. I haven't seen it for years. Probably caved in, by now."

"What was it like, Bill?"

"Oh, there was a regular mat of ferns and things on the ceiling. There were some big palm leaves, too, three and four feet long. Some of 'em perfect as a picture."

Palm leaves three and four feet long! I had seen palm leaves in the Mesaverde, but they were either small or very incomplete. Only days before States had shown me a section of palm leaf, but it was only a piece of petiole, as big as my two hands. The palm leaf Don Guadagni had collected in the Big Bend country was large, but it was broken across the frond. If I could locate just one perfect palm leaf such as Fogg described . . .

"Where is this Red Mountain Mine?"

Fogg stared at me a long moment. He looked as if he was sorry he had spoken of the matter in the first place.

"It's some old, old workings right south of the States-Hall Mine. But don't you get the notion you're going in there. It's been abandoned about fifteen years. You don't mess around with old mines."

A mine might be abandoned, but couldn't it be re-entered? "Just tell me how to get to the picture room. I'll take all the risks in hunting it," I promised.

Fogg shook his head emphatically. "I will like hell. You keep away from that old mine; you'd just go in there and get yourself killed, likely as not, by falling rock."

A nearby miner spoke up from the shadows. "Don't seem like Bird wants to finish this job, talking about that place. Me, I wouldn't go back in them old rooms for a hundred dollars apiece."

"He just doesn't know the danger, is all," Fogg defended me.

By midweek the miners had cut the track block free on all sides. It rested solidly on Bill Fogg's stout props, awaiting the day of lowering. There remained for me now only the unpleasant task of plastering the under-surface. The tracks, after all this trouble, must be protected against possible damage.

"Jack, I'm going to give you a chance to get away from chiseling rock for a spell," I told Ryan. "I'll need someone to hand me wet burlap strips for a couple of hours. It's going to be sloppy work, but I'll bear the brunt of the mess."

I made myself a coat of burlap bags, and put on an old pair of wash pants and a worn-out cap. I had bought a pair of goggles from Cedaredge the day before, and I donned these. I showed Jack how to cut burlap bandages. We had brought a sack of

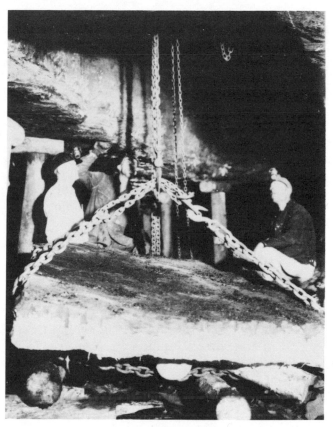

Lowering a portion of the dinosaur trackway slab from the roof of the States Mine. Photo by Bird.

Bird in the States Mine.

123

plaster into the mine, getting water from the nearest sump hole. Under the great block, I looked up at the gently rounded bottoms of the toes of the two tracks and the long, flat stretch of the intervening surface.

"It's going to be a bit of a trick," I said, "to make the bandages stick."

Jack handed me a bandage dipped in plaster and I slapped it on the ceiling. It fell off. He dipped it again, and this time I made it stick. More plaster fell on my face than stayed in place on the ceiling but the burlap stuck. After repeated slapping, I got another bandage to stick. And another. Fogg came by and grinned about my plaster-covered face.

"You've got a job there," he commented.

I paused with the next plaster bandage in my hand and looked along the slab that was eighteen feet long and six wide.

"You going to take me to the picture room, soon's I'm done here?" I inquired.

Fogg grunted and moved on. Three hours later Jack and I were done. I was a solid mass of half-set plaster to my waist, my face was covered, and the rest of me wasn't much better off.

The next day Fogg began making arrange-
ments for lowering the block. From a sugar factory near Delta he borrowed a large chain hoist. Moved into the mine on a mine car, it looked big and strong enough to lift a locomotive or pin down a battleship. Fogg suspended it from a heavy timber set across the roof of the chamber above. The men had chiselled away the upper side of the sandstone slab down to a thickness of about ten inches, which was about what we had expected it to be in the first place. Now they were engaged in making crosscuts through the block so that it could be handled in three square sections. It would be necessary to build a special mine car to get the blocks out, but by far the worst part of the project was done.

"We'll have this job wrapped up in a day or two," Fogg said. He seemed in good spirits about the project. I seized the opportunity.

"And then, Bill, how about the picture room?"

He grunted and started to hedge again, but I sensed he was weakening. He muttered something unintelligible and started to walk away. Then he paused and turned.

"You stay the hell outa there until I'm ready to go with you."

25

A SOFT RAIN fell Saturday night. Sunday the bright green leaves of the quaking aspen had gone to brighter gold. The freshly washed sky had changed from dusty blue to an almost harsh indigo. Our blocks were out of the mine and already rolling to Rock Springs. As Jack and I walked up to the tipple, the world looked good. We paused at the boiler room door and watched the sun pretty up the entrance and almost make the grey boards of the tipple glisten a bit. Mr. States came out of the office door and greeted us with his usual cheery words and smile. Bill Fogg came around the end of the chute. The lanky foreman was decked out in freshly laundered overalls.

"Bill," I remarked slyly, "you look as if you didn't have a thing in the world to do today."

He grinned. "Not a thing. Nothing but see if you still think I'm going to take you to those old workings."

"Well, I was still hoping . . ."

Fogg shrugged. "Well, if it's got to be, it's got to be." Turning to States, he asked, "Charley, do you think we could find that old room again?"

States already knew of my desire to see the big palm leaves. He stepped into the boiler room, walked to a shelf on which stood several carbide lamps, picked out one with a shiny reflector, one that looked as if it might work without undue tinkering. I lit my one-burner gasoline lantern and Jack Ryan lit his. Fogg picked out a carbide in good condition and occupied himself with reloading it with fresh carbide. States looked us over.

"We can try," he said.

Linden Fogg, Bill's brother, appeared around the corner of the coal chute, and saw our little group.

"So it is going to be the picture room? Hot dog! . . . I hope."

"If Charley here can find it . . . or if there's anything left . . . or we don't get trapped by a fall," Bill said glumly.

Linden sat down on a box by the door, stretched luxuriantly, closed his eyes, leaned back against the wall in comfort. But he felt uncomfortable. He jumped up and went for one of the remaining lamps, smiling at me broadly.

"Well, I never had my head examined, but I guess I should've. Before I have it done, guess I'll go along. Never could stay away from liquor or women. When there's none around, I go for trouble in a mine."

States traded his felt hat for a miner's helmet. We left the boiler room together, up the narrow path leading to the mine. The freshly washed clumps of sage along the way gave off an especially aromatic odor. The scattered pines and juniper and red cedar along the ridge did nearly as well. We all breathed deeply, looked once around, and stepped into the clammy dark.

States led. Age had not dulled his zest for life. I wondered if Bill Fogg hadn't overpainted the danger of this trip into the decrepit old workings of the Red Mountain Mine, a wonder that faded slowly with the passing time and the velvet black miles. No one spoke. We passed the first and second forks, the big room with the fourteen-foot seam of coal.

States was still in the lead when we came to the passage leading in to the old States-Hall Mine. At its end the old man turned right instead of left, away from the familiar. We followed him through a couple of deserted rooms into a long passage, bratticed on the north. We headed south, into the abandoned workings. The passage led farther than our light penetrated; the thick blackness gave way reluctantly as we moved ahead, closed in around and behind us as we moved on. The air was dry and still; no live current, as there was in the main, un-bratticed mine. The odor of old coal that had leaked much of its combustible quality into the dark air was the smell of things long dead. Maybe this is part of the smell of adventure and part of the smell of darkness. I whispered to Jack Ryan, "Here we go at last!"

There was an increase in dust on the floor. Tramway ties and rails had long been removed, and dust filled in their cavities. Entries into mined-out rooms were gaping voids filled to the front with darkness. The dusty coal walls reflected no friendly glints, no sound. Footfalls were muffled, as though we trod on a thick rug. A rising black fog stirred by our feet squeezed the gloom tighter about us. Everywhere a new dark pressed in closer, fought back more stubbornly against the light, pressed against our backs, nudged at our elbows.

The party approached shapeless objects that lay directly in our way, long and irregular slabs of stone fallen from the ceiling. We were at a junction with a cross entry and the widened space, poorly supported, had given way. States and Fogg and young Linden started across the tumbled masses, their lights bobbing. Ryan and I followed. Just beyond the litter was a section of roof still hanging. Fogg and States stopped, looked it over with care, carefully ducked under on tiptoe. Their lamps showed a single timber support beneath, an ancient pine, twisted and strained like a bent old man whose burden was slowly becoming too much to bear, who might crumble at any time.

Fogg, waiting ahead, said in a very soft voice, "Don't touch anything."

I held my breath to duck under the menacing mass. I didn't let my eyes rest on the supporting prop.

"You'd only have to bait that deadfall to catch a rabbit . . . or a bear . . . or even all of us," Ryan said softly.

I studied the ceiling, the only thing clear of dust, for fossils. A fragment of bracken and a long twig showed up on the sandy matrix, the twig a species of conifer with long needles.

"Look, Jack, a sprig of pine!" I said. "S'pose I could cut it down?"

Fogg called back . . . very softly . . . "You ain't seen nothing yet. Come along, and tread easy. Don't stomp your feet."

At the end of a long passage States and Fogg paused at the entrance to a room on our left. It had once been well-timbered, but time and the weight of a mountain rested heavily on the slowly failing props. Every timber catching the glow of our light sagged in a way that would have been heartbreaking had they been alive. Some were badly telescoped, others bent sidewise like the drawn strings of crossbows. Still others had twisted and snapped, and stood there stupidly, only because they were too tired to fall over and because their mates were still carrying the load. Not one wooden old man bent under an intolerable burden, but a whole regiment. States and Fogg tiptoed in, bent their heads to avoid the lowering ceiling. "Don't anybody touch anything," Fogg said softly. "Don't bump your head, for God's sakes."

The telescoped timbers were objects of real curiosity. These were the ones with no structural weaknesses, ones that could get no relief in bending. Their stout fibers had only one way to give; they broke within the mass of the wood itself, the fibers bypassing each other. Their telescoping middles sprouted brooms.

Bill Fogg, treading softly ahead, seemed always to be listening, listening to the silence with the care and attention a true music lover would give to a great orchestra.

I noticed overhead odd protuberances here and there, often overlapping each other, lacking definition, but their identity was plain. "Look, Jack," I whispered. "Dinosaur tracks!" Trampled like a barnyard, overhead was the playground or parade ground or highway of the giants, apparently converging in this area.

Jack shuddered. "Under the circumstances, I don't care if I see here the last of those suckers."

It was a great relief to pass the last distorted timber and enter the comparative safety of a better passage heading south. "This should bring us pretty close to where we want to go, huh, Charley?" Fogg inquired.

"I think it will," States said, after considering a moment. "It was well over on the east side, if I remember."

Again we passed a series of empty, tottering rooms, each more desolate than the last, each as filled with black and brooding darkness. Then dead ahead of us appeared a blank wall, black on black, blocking our way completely, a great mass of fallen rock. My hopes of seeing the great palm leaves sank. Here no mere section of ceiling had collapsed, but a whole room. Mr. States stepped to the foot of the mass, held his lamp high, as though looking for a hole. He motioned to Fogg.

"Here it is. Show your lamp over this way."

There was a hole. We in the rear watched him scramble up into blackness. Fogg scrambled from sight. Then his brother Linden. I was next in line, and as I scrambled up the jumbled pile of rubble Bill gave me a hand. He pulled, and I found those preceding me crouched in the close circle of their lights; it was almost impossible to stand upright. I looked about. We gathered on a jagged mass of fallen rock which in falling had left this empty space. Ahead, another long darkness stretched ahead, to the south. It was just possible to make out the far end of the fall. The room had been thirty or forty feet wide and several hundred feet long.

"Good Lord, what a cave-in," I remarked to Fogg.

He shrugged. "Nice it happened before we got here."

We picked our way, crouching, over the fallen mass. I estimated a hundred thousand tons of rock had tumbled down in this one tremendous collapse.

"The wind from this would have blown out your light from a few hundred yards away," Fogg remarked.

At the far end our procession made its way down off the fall. When we had all straightened the kinks out of our backs, States led us through another room with one end of the ceiling slumped in a formless mass. I began to lose sense of where we were and all sense of direction. We were in a nowhere made of dust and decay and sagging timbers and tumbled rock. We were wandering aimlessly in a lost and lonely underworld under the feet of long-dead giants. States and Fogg trudged on with stolid caution. Would the next room be the picture room? And if so, did it still stand, or was all its wonder shattered on the floor?

"I'm afraid we passed it, Charley," Fogg said. "It can't be this far south."

Mr. States paused atop a huge pillar of fallen rock, flashing the thin beam of his light here and there. "You know, I believe we're too far west," he said quietly. "Let's try over the other way."

Linden spoke. "Now, if you'd ask my opinion, I'd say it's still farther south."

Ryan whispered, "What I want to know is which way's home?"

We walked, climbed, stooped, climbed, stooped, walked. It seemed an age back to morning. I remembered the day outside, pictured the bunkhouse, recalled the austere old tipple resting somnolent in the Sunday sun. The golden leaves of the quaking aspen were out there twinkling in the sun; the breezes fluttered golden leaves up into the azure sky.

"If we don't come across it soon," Mr. States remarked, "we'll have to give up, I'm afraid; I have an appointment outside right after lunch."

Fogg was a long shadow, listening and peering into darkness. "We can't be far off, Charley," he said softly. "It's got to be right around here somewheres. Let's try a little farther north."

We rested, finally, in sheer weariness. Fogg seated himself on a great pile of rubble. I picked a rock close by. The others followed suit. We were in the middle of a caved-in room. Our thought was that a spot where rock had fallen was safer than a spot that hadn't yet had its turn. Presently States rose to leave for the outside world. Fogg, reluctant to move, said, "We'll be along shortly."

States's footsteps died out, and for moments no one said a word. Silence became mother of little silences; the singing of my gasoline lantern only deepened the silence all around. Linden sat, seemingly asleep, his head in his arms. Jack beside him idly toyed with a bit of rock, then tossed it aside. It tinkled down the pile of rubble.

Bill Fogg stiffened, on the verge of leaping from his seat. Then he realized the source of the tinkling sound. He spoke sharply to Ryan.

"Don't ever do that again, in a place like this!"

Ryan looked up, surprised at the sharpness of Fogg's tone.

"It's sound like that gives a man warning. If we'd been under a roof ready to let go, maybe the first tiny warning would let loose a pebble thataway somewheres. A man hearing it fall, he'd have maybe a fifty-fifty chance to get out from under, to jump for an entrance, maybe to save his life."

So that was why Fogg was so quiet, why he was always listening, listening even while he spoke! I learned something. So had Ryan. Ryan had been underground, but only briefly and a good while ago.

Fogg got to his feet. "Well, boys, let's go. Maybe I'll come back and try for the picture room again. No; leave out the maybe. I kind of hate to give up, once we've started. Right now I need a breather. I need another day for a fresh start."

"We'll be ready when you are, won't we, Jack?" I turned to Ryan.

Ryan peered around into the dark. He nodded, but glumly, with a lack of enthusiasm.

"What's the matter, Jack?" I asked. "You developing the same complex as Bill and the rest, about these old diggings? Now? At this stage of the game?"

We ducked our heads to clear a sagging ceiling. The shadows of collapsing timbers moved in and out of the gloom, bent over in agonized silence, sprouting witches' brooms about their middles.

"No, it ain't that," Jack replied. "It's just that I was born on Friday the thirteenth. I never know how far to push my luck."

26

"T HAT'S THE OLD airshaft," Bill Fogg said. "The picture room's got to be close by."

We stood in the center of a small square room in pitch dark but for our lamps, which served only to push the darkness back a little. Far above us there seemed to be a glimmer of light, light of a different color, the color of daylight. We couldn't be sure; we couldn't even be sure it was important. All that mattered was that Bill Fogg had recognized a landmark. He turned confidently and started for a nearby entry.

"It should be right across the next room to the south."

We stepped into the room. Ryan and I held our lanterns high, approaching the opposite portal with Fogg. It was partly choked with great slabs of rock. Fogg swore.

"Goddam place's fallen down."

His voice carried but a trace of the chagrin I felt. We walked over slowly to the pile of fallen rubble.

"I was afraid all along this was what probably happened," Fogg said. "And it sure happened."

We climbed over the rubble into the caved-in room. The litter was five or six feet deep, but here and there it was possible to stand upright on top of the pile. I looked around in despair. Nothing but this pile of rock, an empty place in the darkness, a now meaningless name. But I wondered if it might be possible to salvage something from the meaningless rubble.

"Let's sort over some of the pieces where they spilled through the entry," I suggested to Jack, "and see if we can find any remnants of broken palm leaves."

The pieces of fallen scrap were, for the most part, small and thin-bedded. One of them bore a series of narrow flutes, a section of palm leaf from near the petiole. The impression was perfect, though nothing was left of the carbonized leaf. I showed it to Fogg. He nodded knowingly. "That's part of one of them, all right."

I laid it aside with a new hope. Even now, it might be possible to piece together a few fern leaves. The discovery substantiated Fogg's earlier descriptions and settled beyond doubt the identity of the room.

"One of the nicest palms was over by the far entry," Fogg said. He pointed to the passage on the far side of the room. "Then there was another beauty right about where we stand. And a whopper over there next to the coal rib." It was heart-rending to hear him.

"Then there were ferns and hundreds of other leaves. Some like trees we have now. And they were everywhere. The big palms was the prettiest, and there were more of them than I can remember. The whole roof was like a picture had been painted clean acrost it. Or carved into it."

I squinted my eyes tight shut to see it. The material in this room would have outclassed all the rest of the plant collection Barnum had gathered for the whole season. Then too, the geographical factor of distance would have assured us new Mesaverde types. The scientific value of the lost display was impossible to estimate. If only we could have gotten here sooner . . .

Ryan found another section of palm leaf, only hand size but similar to the fragment I had laid beside my lantern. It didn't fit. We continued searching the rubble, but the leaf that had fallen here was apparently shattered beyond recovery. Traces of smaller leaves were entirely lacking. Nearly everything had been reduced to powder in the seven-foot drop of hundreds of tons of stone. It was as though a hurricane had swept through an exotic garden and left not even the stumps of its beauties. Ryan continued poking among the ruins, but I went to explore the surrounding area with Fogg; maybe the fossil zone had extended beyond the limits of this one room.

We explored the ceiling of the next room north. A few broken edges of leaves close to the coal rib indicated the margin of the great mat of foliage that had lain here on the old Cretaceous swamp. The room to the south had fallen. I trudged gloomily over and through the rubble, exploring the region to the west. Everywhere the ceilings were down. We climbed back into the west end of the picture room. On top of the pile, again approaching the spot where Ryan still worked, I noticed a small remnant of ceiling still hanging precariously in the east end. Supporting this twenty-foot section were two badly telescoped props.

"What's down under there?" I asked Fogg.

"Darned if I know . . I don't remember. This section was away up here where nobody noticed it."

I moved closer and swung my lantern under the hanging rock.

"That don't look any too safe to me!" Fogg warned. "Just hanging there on two rotten props. Might not fall for a year. Then again, if I was to toss you a bait for that deadfall, might catch one of them crazy museum fellers 'fore you could duck outa there. If I was you . . ."

He spoke too late; I was down under the hanging ceiling before he finished speaking. My lamp caught the glint of shining black plant impressions everywhere.

"Bill, there's a *lot* of leaves down here!"

Fogg looked down, as unmoved and unmoving as if carved from granite. I swung my lantern about, gloating over the fantastic display. There were the easily recognized leaves of a fig, with simple indentures on the margins. A large waxberry, spatulate and slightly pointed. Pine needles in thick clusters attached to bits of twigs. Lacy fronds of ferns lay all about, delicate in form.

I moved out under the middle of the ledge. A dark shape was etched across the surface of the stone, partly hidden by a thin skin of matrix, but the visible portion was shiny black, ribbed like the spread of a big fan. A palm leaf, a great black leaf three or four feet long.

"Bill, there's a whole palm leaf here!"

I looked up at Fogg. He didn't say a word, just looked at me dully with a pair of well-jaundiced eyes. It came to me I was standing under tons of rock that might be brought down with a hearty sneeze. Mine would be an easy death, but Bill Fogg would be responsible for it. I could appreciate the foreman's concern, but here was fossil treasure in abundance.

"Bill . . ."

"When are you comin' to hell out from under that rock?" Bill's lips were a tight line.

"Bill, this is exactly what we want. Come down and look at it . . . please!"

He shook his head in disgust and despair. He came down gingerly, like a cat on ice. Maybe the excitement in my voice moved him; maybe he was coming down to brain me. The beam of his carbide light merged with my lantern's light and flooded the palm with glittering radiance. He spoke in a strained, soft voice.

"That's one of 'em, all right. Only, I never remember one this far back in."

"And in good condition and complete. Only, how are we going to get it out?"

Bill stared at me in amazement. He spoke in hushed tones, hushed to keep the sound waves from shaking things down upon us, hushed in amazement at my stupidity. "You . . . you don't think you're goin' to collect this thing!"

He turned for the greater safety above. I caught at his arm and begged. "Can't we? Can't we prop up the ceiling or something?"

"Let me up outa here!" He said in a soft, muffled, undisturbing scream. He leaped for the rubble pile almost ludicrously. He brushed himself off, not from the dust but from the very thought.

I could see his point. To attempt to cut into the heavy block as it poised there precariously would be suicidal. But if it could be retimbered . . . I could shut my eyes and see this magnificent block on display back at the museum in Dinosaur Hall and people all gathered around. Reluctantly I joined Fogg on the rocks above.

"Bill, this job's not impossible. Can't it be retimbered?"

"Y-y-yes-s-s," he conceded grudgingly. "But think of the job just getting timbers down here. Look at the time it takes just to get in and out of these old diggings. Remember, there's no way to get a mine car down in here; how you going to ever get your rock out, huh?"

All his arguments were valid, but not overwhelming. To all, I had one answer, "You just wait until we show this to Dr. Brown!"

"Well, I just can't see . . ."

"When we get the big footprints out of the way, you will see. You'll be working on this job the next day. Well, maybe not until the day after."

Jack Ryan broke in, "Or the day before . . . if you know Brown. He'll have it out."

Next morning Brown came in from Grand Junction. Together we looked over the track slab, now nearly ready for lowering. Then I asked Fogg to go with us into the old workings. In the picture room Brown clambered down under the dangerous overhang without hesitation. His mind and mine ran in the same groove when it came to fossils. We were reasonably careful crossing streets. But whether rocks were stoutly braced or held in place by cobwebs, darkness and inertia was all one to him. I knew exactly what he'd say when I held my lantern up to show him the palm.

"R.T., that's fine! That's a beautiful leaf! We have to have it." He scanned the ceiling, delighted with the other plants. "You should be able to make a fine collection here," he added.

Fogg accepted the verdict without argument. He had again resigned himself to the inevitable; there was nothing to do but to retimber the ceiling and tackle the job.

"As soon as the footprints are out of the mine," Brown added, "you can get right to work on this. I've already arranged for a truck to haul the track slab to Rock Springs. I'll go back by train to supervise loading into a freight car. Then I've got to go on to New York. This'll be your baby, R.T.; give it the best of care." He wished us all luck and left.

In the track room I set up my old five-by-seven camera, and readied flash bulbs to record the lowering. Fogg moved about with the miners, arranging a great chain under the forward section of the slab. Vaughan Nichols pushed a special mine car into position. The car track didn't lie directly under the slab, but Fogg had arranged a lateral chain hoist to swing his burden over it. The mine foreman was in the upper chamber one minute, down in the forest of props the next, checking and rechecking the final details. He gave the order to lift the slab slightly and lent a hand to the two miners on the chains.

The chains rattled and clanked, and the block came slowly up. Ryan and Fogg moved out the loosened supporting props. Fogg eyed the job with satisfaction, then rejoined the men on the chains.

"Well, I guess we can let 'er down now."

Again the massive chains groaned and clanked.

The palm leaf collected by Bird from the "Picture Room" (called the "Tropical Room" by Brown) of the Red Mountain Mine. Photo by Julius Kirschner. Courtesy Department of Library Services, American Museum of Natural History.

The forward end broke from the ceiling, wavered pendulously and uncertainly. Then the whole block settled down and the lowering process began. My first flash picture blinded us all. Then Ryan at the lateral hoist began drawing the block toward the mine car. Fogg leaped to guide it in its final drop. Lights flashed again to immortalize the moment, and the great block settled to rest.

The men came to gather about the block, grinning over a job well done. Roy Aldrich, bare to the waist, clean rivulets of sweat running down his coal-dust-covered chest, was shot topless for the final photo.

"I'd sure like to have one of them pictures, when you get 'em developed," Charley Fogg suggested.

"You'll get pictures all around," I said.

Bill Fogg came back from the far end of the room, reeling in a steel tape. "We'll have to widen the passages a few inches here and there to get her out in the main entry," he said. "But that's a breeze, atop of what we've already done."

The day after, I phoned Brown at Rock Springs. "The blocks are as good as on the way," I told him. "They're not crated yet, but they're out of the mine. Should be loaded on the truck by dark."

"I'll be expecting them," he said, "and make my plans accordingly."

Fogg and I studied the blocks, now lying plaster side down, beside the tipple track. They were still too thick. "We can cut the weight down a good bit, chiseling off a couple more inches," I sug-gested. Fogg gathered three or four men, and they went to work, but it went slowly. We finished one block, reducing its weight about three tons. I went to work making a box for it with one of the men, while the others went to work on the next block. Fogg himself went to the States-Hall tipple for more timbers; it would be necessary to build a heavy tripod for the chain hoist in order to turn the blocks with their partly completed crates.

At four o'clock, normally considered quitting time, two of the great blocks were not yet ready for loading on the waiting truck. The largest block, possibly as a result of chipping off excess stone, developed a crack and had to be moved with some caution. At four-thirty it began to drizzle, which of course brought dark on early. When we turned the cracked block over to spike up the opposite side, the fracture widened. A four-foot mass came loose from one end. So a special box had to be built for the four-foot fragment. We still had the third block to go when the drizzle changed to rain. The miners lit their carbide lights, and the rain hissed merrily on the lights and on the globes of our gasoline lanterns. That was all the merriment there was. We grew wet and hungry, and we all took off to hurry to town for a bite to eat. As we came back the steady rain turned to a ragged downpour. The soft shale around the last block became gumbo. Chains caked with mud choked the pulleys of the hoist. As it grew colder, we didn't notice the wet so much; we noticed the cold.

A HEAVY TRUCK rolled by the bunkhouse slowly. The sound reached my ears but just stirred hazily in my mind. It seemed Ryan and I and Fogg and a couple of good old boys were somewhere struggling in mud and rain. Water fell in sheets and the night was going to run on all night and all day and it was away past midnight and morning was cancelled for the day but the job of loading three great blocks of stone had to be finished. And the last block gave us hell; it had to be lifted high and swung over on top of the other two. Then we were jouncing and shaking along with the driver as far as the good road, up a steep trail treacherous with mud, wheels threatening to bog down in the gumbo-like slime. We'd never make it we'd never make it we'd never . . . and the truck motor outside my window shut off and I opened my eyes. Outside my window a late sun was shining in on me. It fell on the floor and some of it splashed up on my cot.

"So it really all happened," I thought. We were really through with the big footprints. I wondered if Brown would have such trouble getting them unloaded and aboard a freight. I hoped not; good old Barnum was getting along for that sort of stuff.

After breakfast, Jack Ryan and I found Bill Fogg standing in the boiler room door. This lean old boy was tougher than nails. "You two softies look like you could stand a day off," he said. "Tomorrow we'll repair the track on the tipple we tore up getting the blocks across. Probably take us most of the morning."

"About the picture room . . ." I said. I asked about timbers for the job.

"Well, there's what we need lying around the stockpile on the mountain. It'll take about eight. And I've got a plan. I think we can take 'em into the mine through the old escapeway."

"The old escapeway?"

"Yeah. That's a tunnel that used to run from inside the old workings to the outside of the mountain. If we can use that, it'll save us a heap of time and trouble."

The tipple repair was accomplished next morning, and after lunch the party was ready for the mine again. We all piled into the Ford truck, and Ryan drove us up on Red Mountain. Someone had brought along a crosscut saw, and Fogg had an axe. We came to Fogg's pile of timbers in an open place among the trees. We selected eight stout props and continued over the east shoulder of the mountain, following a dim trail. Far below us, we knew, were the old mine workings. The trail carried us through a thick stand of piñon pine and cedar to a steep slope above the great brown cliff of the Rollins sandstone. Fogg raised a hand for Ryan to stop.

"The old escapeway is just under us," he said. "We can slide the timbers down to it from here. But first we'd better have a look. It's been a long time."

We followed him along a neglected path to an opening in the forest, where the bare ground fell away into a wide but shallow depression. "Tunnel's probably pretty well caved in," Fogg grumbled with his usual pessimism.

We climbed and slid down into the depression. It was wide open and funnel-shaped, with a black hole in the center. The hole pitched down sharply into the sandy mountainside. We lit our lamps and lanterns and went on down, slipping and sliding, into cool darkness. Presently the passage leveled and we entered a small chamber, apparently on a plane just above the Rollins sandstone. Our stratigraphic position in the mountain must now be about the same as that of the mine; by following the slight downward dip of the beds, we should be able to walk directly into the old workings. But our progress was barred here by a fall of sand.

Fogg shot the thin beam of his carbide light

above the fall. The light pierced a low passage about the size of a gopher hole. Roots and rootlets of trees and plants drooped down into the narrow passage. Pendulous masses of dirty cobwebs draped the roots.

"Looks like here's where we start to crawl," Fogg announced.

He started in. I followed his threshing heels on hands and knees, pushing my lantern ahead of me. His head and cap and shoulders swept away many of the cobwebs, but he left some for me. At every touch the dangling roots released showers of dry sand. After crawling about thirty feet I stopped to brush away a sticky mat of cobwebs from my hat brim. This caused several frenzied spiders to seek firmer footing on my face. Behind me, Jack Ryan spit out any sand and cobwebs I had missed. Fogg, ahead, crumbled and rumbled along like an irritated groundhog in his burrow. "Don't touch anything that doesn't touch you first," he warned. "Don't touch anything overhead."

After we had wriggled an immeasurable and intolerable distance, the bottom of our gopher hole dropped, the passage widened. We could stand up. After I brushed another ragged veil from my hat brim, I could see the dirt-smeared white of Fogg's face in my light. He glanced about and seemed well pleased; there was another cave-in just ahead, but it didn't seem as large as the one we had just crossed. "If the rest's no worse," he said, "we can get those timbers through here."

We crawled over the next cave-in. And the next. Finally the loose sand top consolidated into a decent roof. Then we caught sight of a coal seam rib. We were in an unfamiliar passage, but the dry odor and the silence told us we were in the old Red Mountain Mine. We came upon a room with a caved-in ceiling; we were back in the wreckage of the old workings. Fogg commented, "I dunno which is worse, coming in through the old States Mine or through this escapeway. But this is a whole heap shorter. Besides, we're here now."

Then we crawled back out. We were, as always, amazed by the reality of sunlight when we came out of the dark, but we noticed the sunlight had a good bit of slant to it, a westerly slant.

"Looks like another shift, to get these timbers in," someone said. We held a consultation on the truck. We agreed that, if we took the timbers into the mine, we'd go ahead and set them up regardless of the time. "If you fellows want to carry it on through," I said, "I'll see you get an extra shift for it, which you will pretty well have coming anyway."

We unloaded the timbers and rolled them down the hill. One by one we dropped them into the gaping entrance and watched them slide out of sight. We followed the last timber and found the lot lying in a crazy pile in front of the first cave-in. "I guess we'll all have to climb into the hole, line up on our backs, and pass the timbers in hand to hand."

We wriggled into position along the tight passage, like sausage links. I heard Ryan stirring at my feet, and I moved my lantern as far to the side as possible, where it still got in the way of my elbow. Ryan continued to make threshing sounds, and then a timber bumped my shins. I couldn't rise up because of the low ceiling, but I managed to double sidewise and grasp the end of the timber, dragging it forward over my prostrate body. I was still able to knock a little sand from the tree roots. Beyond me the end of the timber dropped and began to dig in until Fogg got his hands on it. Another timber bumped my shins.

The lamp at my elbow got hot. So did my elbow. The little air over me became an oven. Oven-dried, these timbers would be, by the time we got the last one in. This went on. And on. A sharp knot jagged my leg. But this was no time for bandages. After a goodly while, Fogg called back, "No more, until I get these off my back. We're choking up the tunnel in here."

Behind me—at my feet, rather—Ryan groaned, "If I ever go off on another museum expedition, I hope I die in a place like this and save on the undertaking bill. And I hope they stuff me and use me for an exhibit."

We crawled ahead, cleared the passage, got all the timbers past the first cave-in, wriggled in to the next. The first timber to slide over my feet was the one with the sharp knot, the one that snagged me before. But this time, in a new and untenderized spot. There was another one with rough bark that oozed pine gum, no matter how well it got smeared with sand. The rest were without bark, which means the knotty protuberances were more sharp-edged.

Finally, groggily, we came to the beginnings of the old workings. Fogg and I, exploring ahead, recognized our position; we were only a short way from the picture room. We gave the timbers their last move and piled them up in the entry on the south side of the room. Bob Campbell still toted his crosscut saw, and Fogg his axe.

The lanky foreman stepped into the room; the next job was his to handle. He swung the beam of his light along the jagged face of the hanging ceiling and into the space below, where the two weight-splintered props stood. The ledge seemed suspended by a hair . . . or less. Held up by sheer inertia; simply too tired to fall. Fogg stepped under it with the careful precision of a cat treading through shallow water.

He dropped his steel tape from the ledge to the floor. I held my lantern to give him the best light I could. Very softly he called out a figure to Campbell. He stepped lightly to another spot and repeated the process. Ryan and the other two miners stood at my back, silent, watchful, on tiptoe. Another spot. Another softly spoken figure. When he was done, he rejoined us, still cat-like.

Campbell sawed a timber to the first measurement. Fogg took his axe to the waste end, about sixteen inches long, and fashioned it into a long wedge. Linden measured off another timber, and the process was repeated until we were out of timbers.

Again Fogg pussyfooted under the hanging rock, carrying a prop and a wedge. The top of the prop lacked about three inches of fitting. Fogg gently pushed the end of the wedge into the space. It caught the ceiling surface and the top of the prop, serving to hold the prop upright. He hadn't brushed a grain of sand from its place.

I watched in fascination. The man under the ledge was no longer simply a lanky miner with lean face; he was an elemental spirit of the air, moving too lightly to disturb soap bubbles. The rock overhead, the props, the wedges, all obeyed his slightest touch. He turned after his seventh prop and went back to tap the end of each wedge with his fist . . . gently. The force exerted against the ceiling was extremely slight, but firm. Fogg went back over the wedges, striking each more forcefully. Then he tapped them each a turn or two with an axe. Light taps. Then again. Until the weight above was absorbed by the props.

"That ought to hold 'er," Fogg finally decided, "if things don't start settling above the ceiling."

I looked at my watch. We had been working nearly sixteen hours. It seemed a long while back to morning when we had started off by repairing the tipple track. When we came out the mouth of the old escapeway, it was a surprise to see the stars were out.

Next morning we didn't discuss wriggling in through the escapeway. The abandoned tunnel had served its purpose in getting the timbers in, and we preferred the longer trip in through the old workings. I carried my camera and equipment. Ryan was loaded down with plaster, burlap and a bucket of water. Bill Fogg carried a coil of rope. In addition to hammers and chisels, the others carried planks, and one carried a few short pieces of pipe to use later as rollers in moving the specimen. We reached the room at nine o'clock.

Fortunately, the section of ceiling carrying the big leaf print was relatively thin-bedded; a bedding plane ran through the rock, four inches above the beautiful print. By cutting a channel through four inches of matrix a block could be outlined that should drop free readily of its own weight. To protect the surface of the print, Jack and I slapped on a plaster jacket. Fogg and the others made up a cradle of planks, and suspended it on ropes between two of the props under the prospective block.

Ryan and the miners started chiseling. I turned to the array of surrounding plants. Splitting off small slabs containing mats of fig leaves and fern fronds proved simple. My work went along rapidly, but the channeling of the big block was tedious and slow. The sandstone was hard, or it wouldn't have been there. Swinging a hammer upward, with head and shoulders bent away back . . . this was hard; it could be endured only for a few minutes at a time. The men took turns, two working and two resting. It was evident the job would take a lot more time than we had hoped and planned.

"I don't like the idea," Fogg said, "but we'll have to use heavier tools. We'll never get through at this rate."

He hadn't wished to disturb, to jar, the ceiling, even propped up as it was, with more than the blows of the lighter hammers. Now the men turned to heavier sledges, then to one of their own coal picks. The chips began to fly, but Fogg was uneasy. Suddenly he held up his hand for silence, and listened.

"Get to the entry, boys," he barked, and bolted to the archway of the nearby door. We leaped after him. The foreman listened again. He strained to listen.

"I thought I heard the roof beginning to work.

I don't like using that heavy pick, but it can't be helped. We just have to go slow, be careful, keep listening."

An hour later, Fogg again thought he heard or felt movement. We all jumped for safety . . . and trudged back. The strain was hard on everybody. Hour after hour passed. The channel was a long wide groove, four inches deep, around a rectangle five feet by two. The block might drop into its cradle at any moment, hopefully not catching anyone. Or it might take forever. We decided it meant to take forever. Either it was not quite cut into the bedding plane we expected to hit, or the transversal cleavage was incomplete. It was midnight.

"Let's knock off and eat, and maybe rest some," Fogg suggested.

The men had all brought big lunches, but, except for snatched bites, we had given little thought to food. We moved to the outer room and found seats on the fallen rubble. I set my lantern on a block as a light for the group. We unlaced the strains we were all under, and the men began to talk.

"Bob, do you remember the time those two fellows got killed over in Tipton Mine?" Fogg asked. "It was just dumb luck old Tipton didn't get his the same day."

Campbell replied, "I don't remember much about that one, but I'll never forget the big fall in Jim Coulter's Mine. Tank Hatterly and Tom Roscell got smashed flatter'n barn doors. It wasn't anything anybody could have helped, either. A sort of freak parting nobody knew was there. Boy, was that a mess! I had to help dig 'em out next day. What a mess!"

Campbell took a bite out of his sandwich and reached for his thermos jug of coffee. "They really should've left 'em where they was. If I ever have that much rock fall on me, just skip the undertaker and fetch out the preacher."

The stories made me wonder what might have happened in this mine. "Tell me, Bill," I asked, "was anybody ever killed here?"

I knew at once I had spoken out of turn. Bill looked at me scornfully. "Nobody's been killed here . . . yet," he said.

Campbell set down his coffee and went on. "Closest call I ever heard tell of in a mine, one time one of Jimmy Simpson's mules kicked a carbide light into a pile of hay in the stable room. Old Jimmy and his two sons was loading cars nearby and they run to see what was on fire when they smelt the smoke. While they was putting it out, the

wall fell, right in front of the rib where they'd been working. All that saved 'em was the mule kicking over that lamp."

"How'd any fool leave a lamp where a mule'd kick it over into a hay pile?" young Linden wanted to know.

"Don't ask me. Anyhow, it was the luckiest thing he ever did in his life."

"Funny thing, it didn't do him much good. He only lived to die a couple years later in Grand Junction in a car wreck. That was when cars was still scarce around here. He sold his horse and buggy and bought a Model T, and then killed hisself with it."

"Seems like he should've stuck to his horses and mules."

"Did you ever hear how Bull Mosely got killed in that mine his father ran?"

The gruesome talk went on until lunch was done. I wondered at times how much was fact, how much was embellished, how much was for my benefit. Mining is dangerous work indeed, and perhaps the best way to deal with danger is to spit in its eye.

Returning to the job under the dangerous overhang, the men deepened the channel until there could be no possible doubt the cut penetrated to the bedding plane at every exposed point. Apparently there must have been a hidden roll in the strata still binding the mass to the main ledge. We pounded wedges into the exposed edges of the bedding plane until they were out of sight. The rock popped and snapped but hung on. Everyone was so jumpy we all were constantly hearing the roof working, constantly leaping for safety. Not knowing what else to do, we splashed water from the plaster pail into the crack, just to see what might happen. What happened was nothing. We sat around and waited a bit. Nothing kept right on happening.

"I got it," someone suddenly said, excitedly. "A watched pot never boils. Let's get the hell outa here and let 'er set."

"That's as good an idea as any," I conceded. "Let's go into another room and let it do whatever it wants in private."

We went into the next room and sat down to doze. We were tempted to go away, to walk out and forget it all until we'd had some sleep. But until the slab was down the miners wouldn't be done with their job. Minutes passed. As we weighed the pros and cons of withdrawal, we heard a snap and pop from the next room, followed by the sound of a lunging weight dropping into its cradle.

"She's free, boys," Fogg started for the room.

We lowered the cradle to the floor. I checked on the time. It was three in the morning. Wordlessly, we turned toward the north entry for the long walk out of the old workings.

Ryan and I slept far into the next day. Early in the afternoon we started back into the mine, loaded down with lunches, planks, another pail of water for the plaster. It would be best, I decided, to protect both sides of our slab with a plaster jacket. For the first time, Jack and I went in alone; we knew the way now as well as the mine foreman. The picture room was a lonely spot, without the boys and Bill Fogg.

When we were done with the plaster job, the piece weighed about three or four hundred pounds. We left it on Fogg's flat cradle and laid out a plank track over the rubble of the room. With pieces of pipe serving as rollers, we pushed the block out from under the menacing overhang into the safety of the north entry.

"Now we can leave this place, once and for all," Jack said, gleefully.

We gathered up our tools and looked about. When we drew out our plastered slab and left this place, we'd be the last humans ever to see this spot. In a few years, perhaps only months, Red Mountain

would hunker down and fill in this tiny void like a setting hen on her eggs. There would be nothing left. I took a last look around. A few beautiful little plant impressions remained in the overhang.

"Jack," I said, "you don't know it, but when you get back to the museum from a place like this, you find that what you brought back fills only a little tray or two. And you remember the things you left behind and you're sore at yourself for not having grabbed onto more when you had the chance . . ."

I reached for my hammer and chisel. "Only an hour or two more," I said, "and we'll make a clean sweep."

"Suppose the roof starts working again, like Fogg thought it did last night?" Jack asked.

"Oh, I hardly think my little tapping will bring down the mountain. Remember, we were getting in some hard licks last night."

Jack shook his head disapprovingly but went along. We walked in among Fogg's props and worked for some time in silence. Ryan wrapped and labelled the pieces as I chipped them free of the ceiling. A sizable block of fine fern caught my eye, a block I had tried to split off last night and had failed to free. I'd try it again; perhaps freeing the big block might have done something for it.

But again the block proved stubborn. I hunted up my heaviest hammer and sank a larger wedge

The dinosaur trackway from the roof of the States Mine. A line connects the two footprints believed by Brown to represent successive tracks of the dinosaur (that is, a single pace); the distance between these two tracks is 15′ 2″. Photo by Julius Kirschner. Courtesy Department of Library Services, American Museum of Natural History.

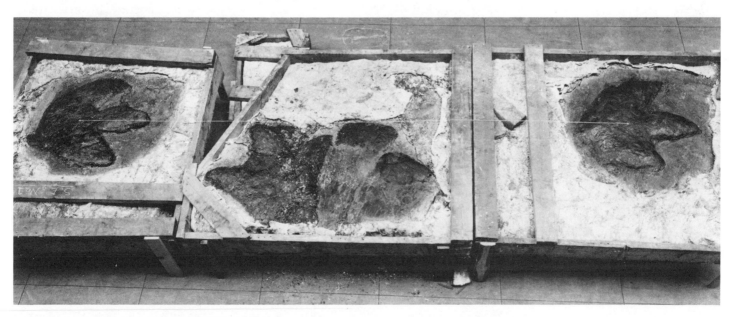

into the bedding plane. I went on hammering, harder; the sound of my hammering rang through the mine. Suddenly consternation swept Ryan's face; he raised a hand sharply. "Listen!" he hissed.

I held the hammer poised. I listened, but only to the hiss of the lantern; the mine was silent as any tomb.

"This roof overhead is working again," Ryan whispered. "I swear . . . I heard it. Let's get outa here!"

"Ah, just your imagination," I said. "Or Campbell's or Fogg's stories. I heard that kind of noise in my sleep all last night."

"I know I heard the roof working," Ryan insisted. "This place is going to be down in ten minutes. I mean it. Let's get outa here!"

I saw he was positive; definitely, he was sure he had heard something. Still, I was reluctant to leave those ferns; they were the choicest of their type. I spent a long minute, listening. The old workings were more deathly still than normally. Then I heard a low, thumping sound . . . which I identified as my beating heart. I shuffled my feet for better footing; I'd rather listen to a beating hammer. I laid on one heavy blow. Jack stiffened, and his coal-smeared face turned chalky white.

"For God's sake, listen!" he hissed.

A horrible, grinding tremor, half shock, half sound, rumbled within the mass of rock above. It trailed off into the mountain, like tired old thunder, and the dull sound was like the break-up of an ice field. I stood frozen in my tracks, but Jack wheeled toward the north entry.

"Jump, dammit, jump!"

I tried to snap into a mighty leap, but my legs froze up. Fresh sound came from above. A great crackling. I was in the middle of a frozen nightmare, a mesh of immobility. Then my legs let loose, and I was flying after Ryan. A mass of falling matter hit my shoulder, spun me, threw me against Ryan. In a terrific clattering crash we fell together into the entry. Flying hunks of darkness bounced and clattered about us.

"Thank God, we moved the specimen!" I thought.

The dust and sound that filled the air gradually settled. Ryan and I lay almost atop the big palm leaf. An unsettled bit of rock rolled down to a better resting place. I tried out a leg; it worked. My right shoulder stung sharply. After a long moment, we disentangled ourselves and got up. Still alive.

The mine was deathly quiet. So were we. We looked into the picture room.

Fogg's timbers were still in position; the higher ceiling hadn't fallen. The general wreckage of the room looked about the same, but a large black mass lay next to the entry. It covered the spot where our plaster bucket stood.

"The coal rib fell, that's all," Jack pointed out.

He was right. A section of the pillared coal vein had crumbled almost in our faces. The settling rock of Red Mountain, undoubtedly affected by our hammering activities, had made adjustments. At the same time it compressed the nearby coal pillar and a section had buckled and fallen outward.

"We sure tried to run into getting killed," Jack pointed out. "If we'd stayed where we were, we'd have been all right." He gave a shaky laugh. "That's the way the influence of my birthday works."

We dug into the black rubble for the plaster bucket. Fortunately we no longer needed it; it was as flat as if run over by a truck. I walked under Fogg's propped ceiling. The fern block that had brought this on had fallen . . . undamaged. So Red Mountain's shake-up hadn't been for nothing, after all.

We finished collecting and turned to the big plaster block in the entry. The only way to take it out of the mine was through the old workings. I dreaded to think of the distance. We piled our tools and plant collections on the block, relaid some of our planks, and began to move the length of the room ahead. Part of the ceiling had fallen. The broken face, the end we must pass under, hung a foot or so from the parting above. I moved the lanterns to a favorable spot, and we inched the block along, a foot or so every now and then. We were careful not to place our planks too near a shattered prop, careful not to touch the crumbling posts. It took us a bit of time to negotiate the length of the room.

That was the easy part. The hard part was building ramps over fallen lumps of ceiling that blocked us here and there. The silence hurt the ears. Then, resting atop one of our rubble islands, we felt a slight movement in the dry, still air. A slight tremor shook the rock we sat on. A dull, muffled thunder came from afar off, out of the dark, off in the distance, from some lost place in the old mountain's innards. A mass of rock had fallen . . . away off . . . somewhere. Jack and I shivered a little; it was much spookier than when it was right on our backs. Then Red Mountain lapsed back into a silence that made our ears ache even worse.

Randomly, at varied and completely unpredictable intervals, lumps of rock had dropped from ceiling to floor during all this operation, as lumps of rock have ever done in old mines. Rock falling on rock is a matter of no moment in the out-of-doors. But in the underground, if it happens well away, this creates a spooky thunder that rolls and rolls. It is no ordinary thunder; it is thunder's ghost. It chills the guts, halts the heartbeat, causes the soul to hunt cover. It rumbles deep inside the hearer long after the last measurable sound wave has died away. But every miner has lived with this sound, randomly spaced and from various distances, all his years underground.

Three hours later we moved into the area where the dinosaurs had made a playground out of our roof. The long room with the badly splintered props materialized out of the velvet gloom. We laid our planks with greatest care across the dusty floor. Inch by inch we edged and eased along; we didn't touch a thing on either side. We shivered when the rollers squeaked or squealed. The footprint-pitted ceiling floated overhead. There was no possible way to avoid this room, where we could almost feel the great dinosaurs stirring in the mountain above us, clumping about in silent ghoulish glee. We were no longer field collectors for a big museum but alien trespassers caught in the Age of Reptiles in a man-made trap. There was no more Red Mountain, no more worn-out coal workings . . . only an eternity of darkness with trampling monsters seeking to collapse their violated world upon our heads.

I shifted one of the lanterns to light our progress. Ryan began to relay the planks.

THE WALL WILL NEVER support all three of those heavy blocks," Mr. Johnson said.

The engineer in charge of installing large exhibits in the American Museum of Natural History shook his head. The huge track slab from the States-Hall Mine was in New York, but the problem of displaying it was still unsolved. An upright wall panel was being considered.

"Well, we could cast the two upper sections in plaster," Brown suggested, "and retain only the bottom original."

The casting required weeks. Bob Chaffee and Jerry Walsh handled the job. I knew what it was like to wrestle with the three-ton blocks, so I was quite pleased that preparation of the big palm leaf took up my time and attention.

The replicas came from the moulds in plaster dyed the color of the original rock. When all three sections were up and joined, and joints along the contacting edges smoothed with dyed plaster, the illusion was perfect. This section of transplanted mine roof, the two great tracks showing the fifteen-foot stride of the mystery dinosaur, greeted the visitor at the east end of Cretaceous Dinosaur Hall. It made a fine addition to Brown's track collection. And the bottom block, original stone, solid and real, resisted the finger nails and pen knives of the public.

The shining black frond of the big palm leaf was set in a wall case next to the head and neck of the biggest Howe Quarry sauropod, where it would always be to me reminder of an unforgettable adventure. Every once in a while I found an excuse to walk past the case. Now the 1937 trip was part of the past, I was thankful none of us had come to harm in that mine. But standing before this magnificent shining black fan I knew I would run the same risks again if necessary, to gather in such a superb specimen. The black palm leaf held a special place in my memory bank too because once, standing there looking at it with me when it was being set up in its case, Barnum Brown told me, "R.T., there's hardly a fossil you can't go out in the field and bring home as well as I."

I owed more than I could ever pay to that man. Under Brown I found myself accepted by the circle in which he and my father moved. Dr. Walter Granger put in my name for the Explorer's Club. The New York Academy of Sciences carried me as a student member from the days when I had attended Gregory's classes in comparative anatomy. My geology was somewhat catch-as-catch-can, but under Barnum I had been helped to grasp its rudiments as applied to fossil-garnering.

The spring of 1938 warmed to new life the dormant trees of Central Park and the museum grounds. Concurrently, it brought back the old itch; each day it grew harder to go inside to work. Getting bones dressed up for the public wasn't near as much fun as finding them.

"We'll get away shortly, R.T.," Barnum told me, sensing my restlessness. There was no money for a big expedition, but he and I planned to spend the summer on the Cloverly in Montana.

The warmth of June and the heat of July and the postponements growing out of moving from inside job to inside job made things more and more intolerable. I was becoming a paycheck-worker going over and over from the fifteenth to the thirtieth and back. With each paycheck, two more weeks of the hunting season had gone to pot. Desperately, I approached Barnum with a proposition.

"What do you say I take the old Buick and get started down on the Crow Indian Reservation?" I suggested. "Maybe I could find something."

He fell in with the idea more easily than I had hoped. "It's been a while since I collected there. More skeletons should be weathered out by now."

He pulled a number of aerial photos showing Cloverly exposures from his desk drawer. He laid his finger on a spot where the edge of a long butte weathered down into highly dissected badlands.

"Right there," he pointed out, "is where Pete Kaisen and I took out five skeletons from five different quarries, after I thought I'd cleaned the place out years before. It's within walking distance from Cashen's Ranch."

He moved his finger to the other side of the butte. "And here's where we collected the armored dinosaur. Undoubtedly there's more material under that butte than we've ever taken out . . . it's like a bank you can go back and draw on every so often."

"I'll leave day after tomorrow," I said.

The Cloverly Formation, early Cretaceous, contained a fragment of the story of reptilian life between the well-known Jurassic era and times like those of the Mesaverde. From this horizon had come Brown's famous pygmy dinosaur. This little carnivore was hardly larger than a turkey, but its skull and jaws in comparison with other dinosaurs were proportionately immense. Any dinosaur found in the Cloverly would be worth collecting. Brown planned to follow shortly.

"George Shea wants to join us," he told me. "I'll write and tell him you're on the way."

A week later George and I, with my younger sister Doris who decided to come along for the ride, set up camp on Cashen's Ranch. The familiar Cloverly colors blazed in bluffs above Beauvais Creek and in the bluff indicated on Barnum's aerial photo. The first day out, we found a camptosaur-like tail running into a bank. The small vertebrae, with long neural spines and attending chevrons, led not to a skeleton but, disappointingly, to the tip of the tail. The next morning I came upon scattered fragments of tendon bones, with a few articulated caudals, but nothing to go with them. George found odds and ends of similar nature. Doris learned what it was to prospect on barren rocks.

Trifling finds continued for several days. We found plenty of rattlesnakes and were fortunate that we always found them first. But old bones were few and far between. We located Brown's old quarries, one by one. Over and over, we would come upon a deep hole or a spreading cavity in the colorful clays, with edges only slightly weathered. There would be old boards and waste ends of plaster bandages lying about, discarded plaster pails, now and then an empty shellac can, rusted from long exposure. The region hadn't had a heavy rain in years; there were no newly weathered fragments to indicate the presence of new and rare Cloverly dinosaurs. A letter from Brown, telling us he'd be out shortly, was the best thing that turned up.

The weeks wore on into August, and Doris wore out and went home. I felt sorry for my kid sister; to her all my days with Barnum were glowing tales, filled with excitement and discovery. Now she had lived with me through one of those long periods of discouragement I hadn't told her about. George and I had nothing to show for all our work but one skeleton without a tail and a tail without a skeleton. It would have been fine if they had matched; they didn't.

We turned to thoughts of new territory. When Brown and I first laid plans, he had said, "And then there's Middle Dome, where I found the pygmy dinosaur. If you finish up at Cashen's, go to Harlowton and look up Al Silberling. Get him to show you the spot where I found the tenontosaur at the dome in 1931. I meant to take it up that season, but there wasn't time."

A tenontosaur was a little plant-eater, akin to the Jurassic camptosaur. In life the light-bodied little dinosaur had been lithe-limbed and probably quite active. An element of slender beauty must have added grace to its movements, as it darted about in quick, jerking steps among the new vegetation of Early Cretaceous time.

George and I found Middle Dome another of those structures like the dome at the Hiawatha oil field. There had been a bending of the earth strata, and a rounded hill marked the apex of the uplift. Surrounding the mile-long rise of ground on all sides were the frayed edges of the weathered Cloverly. Drainage had cut through the ring of circling bluffs at two points, each opening giving access to the valley around the base of the central hill. A ranch house stood on the south side of the hill, but it was empty now; the place had apparently been deserted for years. The spot was ideal for camping, and we moved our cots, mess box and gasoline stove into one of the empty rooms.

Close to the old brown frame dwelling a spring house had been constructed in the clay hillside and walled up with stone. A pool of crystal water several inches deep stood under a shelf of rock at the far end. Two or three green bull-frogs plumped into the water and sought hiding places in the boiling sand. It seemed strange to find frogs here in the desert, so far from every other source of moisture. The water running from the brown Dakota sandstone was cool and sweet.

Our jaded spirits rose with the new camp and surroundings. In the morning Al Silberling came out. Brown's acquaintance was a man familiar with the area and with all fossil-hunters who had worked around Harlowton. He led us confidently up a face in the exposed Cloverly. On a weathered bench of clay we came upon a mound of shoveled earth.

"That pile of dirt marks a place Brown once did some screening," Silberling said. "The skeleton should be close by."

He started to probe with his shovel next to the screenings. Nothing but soft shale. We worked along the face of the hill, investigating all loose dirt except the screenings. The morning wore on. Noon came, and our shovels still had not struck the familiar rasp of bone.

"Well! I thought I could walk right up and find that thing," Silberling said. "But it doesn't look like it now. I doubt even Brown could find it now."

We ate lunch. Our disappointment in not finding Brown's tenontosaur was keen. The little skeleton was undoubtedly somewhere within the Cloverly's colorful clays. The quarry from which the famous pygmy dinosaur had come lay in plain sight as a cut in a ledge a few hundred feet to the east. Piles of screenings at the base of the hill showed where Brown had come back in search of a missing fragment from the femur of the ferocious little carnivore. Brown, with his usual persistence, had retrieved the tiny three-quarter-inch piece. It seemed ridiculous we couldn't locate a whole skeleton on a marked spot.

Silberling mentioned an unexcavated mosasaur. In the afternoon he drove with us outside the dome. At a bend in Mud Creek we paused before exposures of black gumbos and shales. I recognized in the outcrop a marine Cretaceous deposit. Silberling called the formation Bearpaw. Our guide

Painting of *Platecarpus*, a mosasaur, by Eleanor M. Kish from *A Vanished World: The Dinosaurs of Western Canada* by Dale A. Russell (Ottawa, 1977). Reproduced courtesy of the National Museum of Natural Sciences, National Museums of Canada.

stepped across the creek to the side of a blank rise. There were no signs of bones in sight.

Silberling explained, "I've picked up all the surface fragments and taken up some of the caudal vertebrae. If you and Brown ever want to collect this specimen, you're welcome to all I have."

I tried to visualize the mosasaur or river lizard buried deep in the dark clay. The great reptilian head would be several feet in length. The body would be large, the forefeet developed into powerful flippers for swimming. The tail would be long, the whole creature running to perhaps forty feet. This sea-serpent monster, in quest of fish, had thrashed his way through warm Cretaceous waters, along the shores on which had roamed the dinosaurs. I wanted to get my tools and start digging, but the river lizard here was not for George and me. The specimen would probably be a known genus and species, and the museum already had a fine mosasaur on display. The very size of the creature placed it beyond our means for this season.

We went back to the old ranch house, sat on the porch and ate. I asked Silberling about the history of the brown frame building.

"It was built about forty years ago," he said, "by an Englishman named Joe Middecomb, who came here to raise sheep. Sheep-raising proved a failure, and Middecomb moved away, but his loss proved to be other men's gain. This old house always makes a fine camp for fossil hunters. And there's been a bunch of 'em here, too. They've all camped in the house and drunk from that fine old spring. Not only Barnum Brown . . . there's been Hatcher, Stanton, Knowlton . . ."

"Hold it!" I begged. "I'd like to get those names down for my diary. Please."

Silberling laughed. "Well, I'll give you as many as I can."

I put down the first four, and began to add the others: "Jepsen, Farr, Case, Stone, Campbell, Kelver, Simpson, Granger . . . don't forget him . . . Thomson, Arnold, Perry, Harbick . . ."

The list was amazing. Some of the names I recognized as those of real old timers, among the first to prospect for bones anywhere in Montana. Some were unknown to me. Not all, of course, were bone hunters. Certainly some were oil geologists interested in the dome for other reasons. The deserted homestead took on a new atmosphere; when I went back to the spring again, it seemed I could see in the water the faces of all these others who had shared this spot as a base while scouting the surrounding hills. Truly the green bullfrogs and their fathers and grandfathers had known a varied and distinguished company.

In the morning George and I decided to have one final fling looking for Barnum's tenontosaur. "It's bound to be close to where Silberling thought it was," I insisted. "Let's broom off everything near that pile of old screenings. Let's look into the screenings too, just for luck."

We climbed back to the site. George was first to stick his shovel into the screenings.

"What the . . . ? There's a layer of newspapers here!" he exclaimed.

We raked aside the loose pile of clay and lifted the papers. They were dry and in good condition, and at the top of one page the date June 23, 1931 was plainly discernible. A pile of dry bones lay beneath the protective layer of old papers. Broken vertebrae. A fragment of a limb bone. Odds and ends and other pieces. The material was sound and firm, and adhering to it here and there were lumps of red sandstone.

"Lucky we didn't stumble over this," Al said. "We might have broken a bone!"

But where was the bone outcrop? These fragments hadn't been dug from clay. I looked up the slope of the bluff. Thirty or forty feet almost directly above was a narrow ledge of red sandstone. It was neither extensive nor heavy, but on the face of the slope, across a portion of the weathered edge, lay a long, flat rock. Bone protruded from the hard red matrix. I tried the limb bone fragment for contacts. It fit. Here was the rest of Barnum's long-neglected tenontosaur.

We began excavating. The sandstone ran into the bank, and the bank was ideal: a bone-bank with withdrawal privileges. Then our picks ran into soft, silent clay; the sandstone ran out. We dug further, responsive to an old common law of Brown's: always dig three feet beyond the specimen. This three feet gave us a good yard of poor pottery clay but no bones. We searched the length and breadth of uncovered rock, but the sandstone tomb was too small ever to have held an entire skeleton. At least, we had found what was to be found of the tenontosaur.

As the days wore on and the summer season unraveled, Brown wrote me to go south to the Big Horn Basin and work on the Cloverly around Sheep Mountain. In Billings Shea had to leave me. I was sorry he had to go, and accepted the fact Barnum wasn't going to make it out this season.

The smoke of Greybull, Wyoming, showed up in the clear skies ahead. Sheep Mountain was the great lump that arose to the left of the road. Across a few miles of rough terrain were beds I had first seen from the air the summer of the Howe Quarry dig. The area would be hard to reach and had probably never been prospected by a dinosaur hunter. The roads into it were only sheep wagon trails. I was elated that Brown had delegated such virgin territory to me; here, without a doubt, I must find Cretaceous dinosaurs I could take up unassisted.

That was then. In early October I was busily turning over dirt around Shell. A few scattered bones around Sheep Mountain. A partial skeleton of a good-sized fellow in a heavy ledge of Dakota sandstone. The weathered humerus was that of a sauropod, one of the great quadrupeds. Bones of such a creature in this horizon were unusual, but the size of the animal made it impossible for one man to handle. The odd surface fragments, I stored at Paton Ranch.

New suggestions came in from Brown. I photographed the work going on at Dinosaur National Monument. In southern Utah, near the little town of Hurricane, I ran down a lead on a tremendous assortment of mammal bones, found in a cave. The bones were there. Rabbit bones. At Cameron and Chinle, Arizona, I tried again to discover undiscovered Triassic dinosaurs. No luck. Where was the old Barnum Brown touch? If it were ever to rub off on me, any of it, it should have by now.

November 9 I drove into Pueblo, Colorado. Strange footprints had been reported to Brown from here, by a John Stuart MacLary. He had suggested I check them out.

I found MacLary, a young man confined to his bed. Mr. and Mrs. Myers, close friends of the invalid, volunteered to act as guides. "We've been excited about these tracks," Mrs. Myers said. "John was so hoping someone from the museum would come out to look at them."

The site proved to be in Lower Cretaceous. The Myers car drew up on a bench of clay above the limestone floor of Purgatory River. The hard, flat bed was only partly covered by water, and leading into the edge of the thin current I saw large, rounded impressions. The tracks were spaced like those of a biped, the great creature having stepped along in great eight-foot strides. The circular prints,

each roughly two feet across, showed no marks of claws. By what animal could they possibly have been made?

I studied the big tracks carefully, photographed them from different angles. The size of the impressions and the length of the stride suggested a creature of great bulk and size. The trail stretched from bank to bank. Here had tramped no great three-toed *Iguanodon*; here was a massive lumbering monster walking on feet that were round and clawless. I couldn't associate these with any tracks from the entire Age of Reptiles. Were they those of a sauropod, a brontosaur-type? I wanted to believe they were. So far, tracks of these giants were unknown. But the brontosaur was a quadruped, and the sequence here was not that of a four-footed animal. And sauropod rear feet would have to show claws. And quadrupeds were noted for messing up their front prints with their rear feet. I was both intrigued and baffled.

"I'm sorry," I had to tell the Myers. "I'd love to tell you what made this trail, but I can't. There's no doubt it was a dinosaur, but I can't even tell you if he was learning to walk on his hands."

The next morning I got to Gallup, New Mexico, still haunted by the Purgatory footprints. It was a relief to be at the routine job assigned to me here; I was to crate a large fossil cycad, discovered near the city. Brown had seen the bole of this Cretaceous plant some years before and pronounced it one of the largest known. The finder's daughter, Mrs. Seymour, was expecting me.

The bole was lying in the yard. The mass of stone was three feet long, shaped like a beer keg. The surface showed spots where the branching feathery leaves had been attached to the fat stem. I estimated the entire mass weighed better than a quarter-ton. Mrs. Seymour kindly gave me permission to crate the fossil and take it away.

The burly drayman who drove up to haul the cycad away looked curiously between the slats. "That's quite a piece of fossil you got there, mister," he commented. "You do this sort of thing everyday?"

"Only when I'm lucky enough to find them," I said. "I have dry days . . . had a lot of them lately. This is just a fossil cycad donated to the museum."

"Cycad, huh?"

"Something like a palm or fern." I explained how once they had grown, in a warmer world, almost to the polar regions on the north. "All that are found nowadays are in the tropics and subtropics," I concluded.

"Well, I don't know nothin' about such things." He lighted his pipe and took a long drag on it. "Talk about fossils, though . . . you oughta drive by Jack Hill's and see what's in his store window. Hill has an Indian trading post t'other side o' town. Got some o' the damnedest tracks in stone ever I seen. Look like they was made by a man maybe twelve feet high."

A man twelve feet high. "There were giants in those days." Always, in this business, we are hearing about wonders . . . somewhere else; not here . . . afar off . . . I heard about it from a fella . . . Flakes of snow settled down as we unloaded the cycad on the freight platform.

"If I have time before I leave, I'll go see what your Jack Hill has. Thanks." As I started back, the sky broke open and snow feathers came tumbling down in earnest. The aging Buick's side-curtains slatted in a rising gale. I recalled what Don Guadagni once said, "If we had a couple of fish nets, we'd get more out of them than out of these curtains."

I checked the map for a road out. Checked the time. Checked the wind and snow. Too nasty and late to hit the road; too early for bed. So I checked in at a cheap hotel for the night. Cheap hotels are no fun to lounge around in. The desk clerk told me the Indian trading post was only a short walk across town. In those days, Gallup was a short town.

Purgatory River tracks.

145

29

SNOW SWIRLED down the empty street ahead of me, swirled around me like wind-blown thistledown. The Hill place was a barn-like structure between two nondescript buildings, half-hidden by them, half-hidden by the snow. On each side of the door were small showcase windows. I peered in. The window was bare but for an antiquated sewing machine and two blocks of stone. Each block, about eighteen inches long, bore a man-like footprint of giant size. The impressions were completely natural in the way the toes were splayed, in the human shape of the heel, in the clearness of the outline apparently once made in soft mud. A shift in the wind brought a curtain of snow between me and Jack Hill's window. I didn't walk in the door; I walked on down the street. I wanted a minute to think it over.

Could the great five-toed prints be those of a Pleistocene bear? Some fine fossil mammal tracks were known in this part of the United States, but so far as I knew, no fossil bear tracks had been found. There had been Tertiary camel tracks, deer tracks, tracks of a big cat like a puma. Perhaps . . . perhaps . . . the pair in the window belonged to a great Ice Age bear.

I retraced my steps. Snow beat upon the glass, half hiding the blocks from view. I turned the knob on the door and went in. The store was full of Indians. Several fat squaws sat about on the floor, wrapped in their blankets. At the back, around a pot-bellied stove, stood a dozen or so blanketed braves, while others lounged around the counter. The group had apparently drifted in to escape the vile weather. I picked my way toward the open bay of the window.

The blocks beside the aged sewing machine were easy to pull to the back of the bay. The odd impressions looked as astoundingly real as they had from outside. The matrix in which the tracks were cast was a light grey calcareous sandy limestone. I realized that whoever had quarried the rock had, at the same time, destroyed proof of the length of stride. How much better it would have been to quarry them in one piece.

A stirring beside me turned out to be a big Hopi Indian. He too had stepped through the seated squaws to look at the objects that held my attention. He hunched his blanket about his shoulders, and his eyes twinkled.

"Zuni tracks," he grunted. Evidently the big Hopi looked on the Zuni with ridicule. I ran my finger around the fifteen-inch track on one block with, first, disappointment. No claws, therefore no bear prints. Only a man could have made these prints. Only, there were no men with feet like this.

A clerk behind the counter, noticing me, came to the front of the store. "Ever see anything like that before?" he asked with a grin.

I shook my head. "Who's been carving those things out of stone to fool people?" I asked.

"I don't know much about 'em. All I know: they come from Glen Rose, Texas. Mr. Hill brought them west a couple weeks ago, with some dinosaur footprints."

Dinosaur footprints? "Where are the dinosaur footprints?" I asked the clerk. I learned they were in nearby Lupten. I looked again at the specimens before me, looking for signs of sculpturing in the long footmark. The rough surface seemed old and weathered, as though the artist might have etched away any marks with acid. Or perhaps the limestone had been green, freshly uncovered from a well-buried and still soft stratum. Definitely, they were not footprints made in mud and turned to stone; they were clearly sculptured copies of the bottom of an imaginary human foot. A real foot could not make a foot of such clarity in real mud.

Were the dinosaur footprints fake, too? In all my fossil-hunting I had never found a fake dinosaur track. I was curious to see what I'd find in Jack Hill's other store.

It was dark and still snowing when I drove into Lupten. Jack Hill was behind his counter. Four dinosaur tracks lay over in a corner of the store, on separate blocks. By the uncertain light of a lantern held by the owner, they appeared to be prints of a medium-sized carnivore, but there was something too suspiciously perfect about the three-toed outlines. The matrix was exactly the same as that of the human-looking prints in Gallup. The claws of the prints were too perfect. When a beast pulls a claw out of mud, he doesn't leave a perfect copy; the mud slurps about a bit. The fact that man appeared something better than sixty million years after the dinosaurs left stamped "Fraud" on all this artists' work.

Headed homeward next morning, I had opportunity to rationalize the thesis that the longest way around is the shortest way home. Snow was two feet deep in Santa Fe. Raton Pass could be at best risky, at worst, impossible. I bore south, toward Texas. I noted that according to the map I would pass fairly close to the town wherein Hill had been duped. Glen Rose was in a belt of Cretaceous rocks known to carry dinosaur tracks, and Barnum was always looking for tracks he didn't have. It might be worthwhile to look the place over, to see if there might be tracks the museum could use. Barnum, who had finally gotten away, was now in Alberta; there was no one waiting for me. So, when I got to Fort Worth, I turned south. Glen Rose was sixty-five miles away.

Little Glen Rose, county seat of the second smallest county in Texas. Shady little town on the edge of the hill country. A central square, real southern style, with a neat little white limestone courthouse shaded by cottonwoods. A bandstand stood in the parking area by the courthouse. I parked the Buick and went over to look over the decorations ornamenting the stone base of the bandstand. Inset in pieces of the petrified wood that is commonplace around Glen Rose as building material was a fine, large carnivorous dinosaur footprint. I approached a man seated on a nearby bench.

"Can you tell me where I can see more of those footprints in the native rock?" I inquired.

He thought a moment. "Well, they used to be lots of 'em hereabouts, if you knowed where to look. They're all along the Paluxy River in places. Tell you what you do: you drive up the river road to Jim Ryals's. He's got his name on the mailbox. Used to be tracks up by his place; he's quarried a heap of 'em for rock gardens and such. Ryals's place is about four miles up, just this side the fourth crossing."

The river road was graded dirt and narrow. At the first bend the road plunged into the water alongside a broad sandstone shelf. The next time the road turned we waded the river again, a bit deeper this time, but there was a concrete apron spanning the stream. I wondered about the third crossing, but it was no deeper than the first one.

The name "Ryals" on the mailbox in front of a small frame house popped up just in sight of the fourth crossing. No one was in sight, but I heard the sound of chopping out behind the barn. I walked out between the young cedars in the yard and came upon a husky woman wielding the axe, cutting fence posts. As I introduced myself to Mrs. Ryals, Jim Ryals emerged from a nearby clump of cedars, a hard, wiry little man. Children of assorted ages trooped in from the surrounding shrubbery. When I inquired about dinosaur tracks in the Paluxy, everybody was voluble.

"I s'pose I've fooled more with them things than any man on this river," Ryals told me. "If you'd come when the water was down, I could show you some just below the house. Now, we'll have to go below the bend."

We left the woodlot, climbed a fence, and started for the bend in the river. Ryals told a lengthy tale of his experience in quarrying tracks.

"I've had a heap o' fun at it," he said. "Don't put much food on the table, but then . . . what does? Hereabouts, 'bout the only money-makin' jobs is cuttin' cedar posts, bootleggin', and quarryin' dinosaur footprints. And the other two is hot, hard work. Cuttin' out dino tracks is hard enough, but it's a heap cooler. I allers put iron bands around 'em to keep 'em together. But Lord, folks don't want to pay what they're worth."

We came to the brink of a pool, where Ryals pointed out a ledge three feet deep. "My wife and I taken one out of there," he said, pointing to a circular depression in the bottom of the pool. "But that was when the river was down some."

We walked along a bar at the river's edge. "Too bad I can't show you what was along here," Jim said. "They was all along here, but they ripped out in a big rise. When the ole Paluxy floods, she really

Footprint of a carnivorous dinosaur incorporated, with pieces of petrified wood, into the bandstand on the town square of Glen Rose, Texas. Photo by James O. Farlow.

floods. When the water went down, the tracks was gone and this gravel bar was left. Whilst she was up, the water was in the tops of them trees."

"What did the tracks look like?"

"Some was rounded-like. Others was bigger and longer."

Ryals indicated by extended hands tracks of gigantic size. He seemed to take it for granted that I would know what they had been.

"Any claws? Did they show any signs of claws?"

"The big ones did. The littler, rounded ones, no."

What could Ryals have found? I remembered the big rounded impressions on the Purgatory. How I had wanted to see in them tracks of the largest creature that ever walked the earth! But the mysterious prints hadn't presented the proper sequence for the trail of a quadruped, and there had been none showing marks of claws. Ryals's description filled me with excitement, but I tried to keep my voice calm.

"Are you sure there are no more of those tracks left around here?"

Ryals was positive they were all gone.

"Well, have you ever seen any like them in other places on the river?"

He shook his head sidewise; my heart sank.

"Did you ever take any of the round ones up, before the river took them away?"

He had taken up one, one of the smaller rounded ones, but he had sold it to a man in another county. We came back to the fourth crossing, and saw one or two more carnivore trails. Back at the house, we had a cool drink from an artesian well. I left the farm with conviction that my Glen Rose detour had a double purpose.

Next morning I drove out to talk about the tracks to other people along the way. The first two had seen the carnivore trails and had, like Ryals, quarried single tracks from time to time. Like Ryals had said, bootlegging and chopping cedar posts is hot work. One imaginative old gentleman had two lumps of "dinosaur food" in his front yard. He showed them to me proudly, as one scientist to another. They were stone concretions, a common phenomenon, but rare hereabouts. He had obviously gone to some trouble to drag them up from the river; each weighed half a ton or more. I left, marveling more at the old man than at his dinosaur food.

Another farmer regaled me with more surprising information but was unable at the time to display his finds. He remembered having seen "horse tracks" and "man tracks" in the Cretaceous limestone. I continued on my rounds, but my interviews furnished no new encouragement. Even the most fanciful had seen no lizard-like footprints over three feet long.

But everyone I talked with was united on one point: see Ernie Adams, "Bull" Adams. The smartest man in town. "Boy, he can tell you just about anything you want to know about fossils, rocks, Indians, arrowheads . . . and he's maybe the smartest lawyer in Texas, too . . ."

I found Adams in Hilda's Cafe. Someone addressed the huge, burly individual by the now familiar name, and I went over and introduced myself. We sat down together for a cup of coffee. My new acquaintance looked like a farmer just in from the hills. His dark flannel shirt had definitely not been put on that morning. His sagging old grey pants showed a prominent rip, his muscular neck was unornamented by necktie. His dark hair, just becoming specked with grey, was flung back in a mop that neither stood nor fell. His eyes twinkled as though he might be laughing at me.

"So you're from the American Museum," he said. "Not often anyone stops here who talks my language."

Ernest Adams was a lawyer, an Oxford graduate, having gone there on a Rhodes scholarship. No one will ever know if he was really the best lawyer in Texas; he distinguished himself at the bar whenever he appeared, but his interests were generally elsewhere. He gave me, at our first meeting, a discourse on flaking flints and quartzes as practiced by the Indians, based on study and practice. Somewhere in a trash dump outside Glen Rose a future archaeologist may be puzzled by a curious flaked little arrowhead made of beautiful blue glass. I watched Bull Adams make a point by picking out a blue pill bottle with a thick bottom, and in what seemed but a few minutes flaking out a magnificent and perfect little arrowhead. He considered making a Folsom point out of it, but decided against it; he tossed it back in the heap with the fragments of the bottle it was made from.

By a fantastically circuitous conversational route, we came back finally to dinosaur tracks. I mentioned the big tracks Ryals had spoken about.

"I believe you can find some of them," Adams said, "if you keep looking. If you'll go up to the second crossing, you'll notice a gravel bar on your right. There used to be a few big tracks there that fit your description."

I lost little time in checking him out. I found the bar, but the river had buried or destroyed all marks in the vicinity. A systematic search was now in order. So I went up the river again, to search the area about Ryals's farm more thoroughly. When I reached the ledge where Ryals had shown me the first carnivore trail, I decided to clean up the area a bit, preparatory to taking its picture.

The exposure was about thirty feet across at the widest point. The course of the trail for a short distance lay close to the overhang of a higher ledge. It was still partly caked with dry silt where Ryals and I had failed to clean completely the space that showed the length of stride. I began throwing the black gumbo aside. Some large pot holes on the outer side of the ledge made good places to toss the excess dirt.

I did all I could with the shovel, and began to clean prints with the whisk broom. At one point a second trail crossed the first, leading straight into the bank. A track of the second trail was partly superimposed on the first. A thin curl of stiff mud had pressed into the bottom track, squeezed by one of the toes of the last dinosaur to pass here. The delicate curl had never fallen; it had turned to stone, like the track itself. The detailed preservation filled me with wonder.

When I thought I had the scene properly prettied up, I shot the picture. I found a piece of shade and settled down to a lonesome lunch seated on a ledge where I could study the uncovered trail. I wanted to stay and work on this old river all winter. Where the second trail ran under the ledge a few well-placed sticks of dynamite in the right places would crack the overlying rock so that it could be easily removed from the shale overburden that covered the tracks. Then I looked up at the trash hanging high in the trees, souvenirs from the Paluxy's last flood. Then the clouds moved in and in minutes the lingering Indian Summer, whose departure was weeks overdue, left immediately. I finished my sandwich in a chilly fall day.

I finished lunch, picked up my small tools, looked about to see if I had forgotten anything. The camera, the prospecting pick, the shovel. A part of the ledge where the shovel lay was still cov-

Bird and carnivorous dinosaur tracks at the Paluxy River near Glen Rose.

ered with silt. "Always dig three feet beyond," Barnum said. Could anything worthwhile still be left buried here?

I walked along the ledge next to the water. A sizable pothole yawned at my feet. It was not a place where Ryals had taken up one of his footprints and seemed only a cavity in the bedrock. If so, it had been there since the days of the dinosaurs. What had caused it? Because I couldn't think of an answer, I probed it with the shovel. It was somewhat filled with muck thrown in from the other trail.

I shoveled dirt off into the stream. The steel blade rang sharply against a margin of the depression. The pothole was wider at one end. Just beyond, the shoveling revealed a strange rounded mass rising above the surface of the ledge. Not just another pile of muck; a neat lump of moulded stone. It looked as if it had been pushed up from the depression when this had all been but a stiff mud. I scraped out the last of the dirt and was amazed that the depression had the shape of a giant lizard-like track. I was standing in a track over three feet long. I dropped the shovel and went to work with the whisk broom. Shortly I uncovered a big toe, tipped with a claw scar. And another toe, larger still, and with a better claw scar. And a third! I moved to the left front of the depression where there was still some fairly hardened muck. Cleaning it out as fast as I could, I found a claw cavity large enough to fit my hand.

Lost in an impossible dream, I stood up, part of a picture never before seen on the earth; I was standing in the right rear footprint of a brontosaur-type creature. The lump of rock ahead of it was the fossil mud furrow ploughed out of the footprint by a giant foot, nearly a hundred million years ago.

I stepped out of the track. Where had the great beast put the foot down next? Twelve feet ahead there seemed to be a similar pothole. I shoveled a wheelbarrow-load of muck from this depression. The final clearing revealed a nearly identical copy of the footprint I had just left. Just ahead of the rolled-out stone lump in front of this great print was a smaller, a bit smaller, a round impression, clawless like the odd prints on the far-off Purgatory. Surely the front footprint of the same great creature.

Prints of the left feet, front and rear, should parallel the prints I had just uncovered. They did. They made a trail six feet wide. In each case, the rear print was barely preceded by the round, toeless depression of the front foot. The creature had advanced along the mud flat in the manner of all quadrupeds, the rear foot often but not always smudging the print of the front foot. The tracks first appeared at the edge of the weathered ledge and ran upstream, moving south, but beyond the first few steps the ledge dipped under water. The trail led along under the surface, the footprint layer smooth and hard, and then it disappeared under a stream bank ledge.

The beast whose footsteps I had found in the Paluxy River lived some millions of years after the *Brontosaurus*, though he was certainly related to that dinosaur. He stood fifteen feet or more at the shoulders, ran out to seventy or eighty feet in length. I could have ducked under his belly.

How could a great big slab of these tracks be lifted from the Paluxy's restless bed, transported to New York, and hoisted up to the upper floor to Dinosaur Hall? Where could it be displayed after I got it there? Who'd pay for it? It must be long enough and wide enough to show the stride. It would be the most spectacular museum display of its kind on earth. I didn't know where to begin thinking about the problem. Brown had told me the mystery dinosaur prints had cost over ten thousand dollars.

"Well," I told myself, "I can at least make plaster copies of one pair; I'll have something to whet Barnum's appetite. Barnum is very good at digesting big problems."

30

IT WAS THE SIXTEENTH day of December 1938 when I brought the museum's venerable Buick in through Holland Tunnel and turned toward home. I felt well-pleased. I had been able to postpone return to the city and with great justification. Discovery of the sauropod tracks in the Paluxy had gone a long way to ameliorate the let-down feeling that often comes with the end of a field trip. It hadn't been a discovery, really, of course; they had been known since man had found his way into the Texas hill country. It was my good luck to recognize them for what they were. Only the three-toed dinosaur tracks had been known, for just under a hundred years, as the calling cards of dinosaurs. As for the great plantigrade sauropod tracks, the Paluxy tracks were the first known in the world.

Barnum was not yet back from Alberta when I got to the museum, but he had been expected for so long that he was bound to show up soon. My two casts had arrived by express and had been laid out and exclaimed over by one and all. I couldn't wait for Barnum to get back, because I knew he would somehow find a way to solve my big problem when he saw them.

"R.T., we simply have to get a set of those, original, not a plaster cast," he said, just before Christmas. It made for a nice Christmas.

But in the spring of 1939, after the Paluxy had been up and down, it wasn't a Texas expedition that Barnum outlined; it was one back to Alberta.

"The Texas thing will have to be put off until funds are available," Brown said. "And in Alberta we'll be doing work for the Sinclair people. The Sinclair people are in the oil business, not the footprint business. Harry Sinclair has been a good old angel to scientists, and when he bought the brontosaur and drove it to work in ads, he did the dinosaur tribe a lot of good. So it's no more than fair that from time to time we give him what he wants."

All in all, the Alberta season was a good trip and good for me. Much of it was spent in collecting from times back of dinosaur time. And a fair bit of it was spent in making the acquaintance of the Canadian dinosaur contingent. Without the lure of the great sauropod tracks away down south in Texas, it would have been a wonderful, valuable, broadening season. As it was, all Alberta was only a restraint on getting back to where my heart and my footprints were.

As usual, I made my way back to New York circuitously. Collecting invertebrates around Moose Dome and Emerald Lake. Down on the Red Deer River I got acquainted with *Styracosaurus*, *Monoclonius*, other assorted Cretaceous dinosaurs. In the fall, a pleasant assignment in the Sweet Grass Hills of Montana. I left Hell Creek for home just in front of the winter's first big snow.

In New York unexpected good news was awaiting me. Brown had nearly completed arrangements for getting at the sauropod tracks. He had negotiations under way with Dr. E. H. Sellards to work in conjunction with Texas Statewide Paleontological Survey and the University of Texas. Material collected would be split between the University and the American Museum.

One drawback: it was now December. The Paluxy was temperamental in the winter, often on into spring. In an open quarry, I might well find myself in a bad way. I mentioned this, but Brown, who hadn't seen the trash deposited high in the trees along the Paluxy's banks, felt if he wrapped up the thing with the Texas powers, we should grab it when we could. "I'm after Harry Sinclair to give us a hand with the museum's end of the thing," he said. With Brown pulling the strings, my worries raveled out a bit.

Then a letter came in from a Frank Davenport

of Wilkes-Barre, telling Brown of the discovery of "the oldest footprints in the world" on the roof of a coal mine there. Would the museum like to come and verify it, bless the find, do the right thing about helping the local folks beat the drum for such a find?

Wilkes-Barre coal was Paleozoic, a hundred million years back of the dinosaurs. It was in Paleozoic time the first amphibians crawled out on the land, indeed the first feet formed. So if they had footprints there, they would be assuredly the world's oldest. Brown was tied up at the office.

"How would you like to go to Wilkes-Barre and look into this thing, R.T.?" he asked.

"Well," I told him, "I can't leave for about an hour. I'll skip lunch."

Before eight o'clock next morning I drove up to the mine tipple with Mr. Davenport. The January night had been cold; the temperature of the grey snow-covered surroundings stood at two degrees above zero. We hurried into the small office next to the giant tipple. I was introduced to the mine foreman, C. A. Ceeling. A reporter from the local press and the state senator were to be out a bit later, but we would look over the site now. Davenport and I were given forms to fill out, absolving the company from any responsibility for our persons while underground. I thought of good old Jack Ryan and was thankful I had been born on a Wednesday. We were each given a miner's helmet equipped with an electric headlamp and a small storage battery for electric power.

"I'll have to ask you for any matches you may have in your pockets," Ceeling told us. "They aren't permitted in the mine because of the danger of igniting inflammable gases."

We had arrived a bit ahead of schedule and must wait for the next down-cage. A bulletin on the wall listed several gases to be found in coal mines, resulting from the dry distillation of plant and animal matter in the coal-forming process, all dangerous to life and limb. I was a bit amazed by the wide choice. In the well-ventilated Colorado mines I had moved in an atmosphere of comparative safety.

Eight o'clock came. Miners trooped out of the office, and Ceeling motioned for me and Davenport to follow them. We gathered at the shaft entrance under the black tipple. With a great rumbling from above, a suspended platform dropped to the level of the ground. The gate to the shaft swung open, and we all stepped onto a rough platform. There were no side rails to separate the riders from the walls of the shaft, but a beam about head-high ran across the top of the cage, from which several rings were suspended. Each miner grabbed a ring with one hand, holding onto his lunch pail with the other. "You'd best grab one of these hand-holds, Mr. Bird," Ceeling said. "We'll be going down pretty fast, all the way to the bottom."

Someone pressed a button. A warning bell rang. A moment of silence, while everyone caught a breath. Then the bottom fell out of the cage. The sides of the shaft became a grey blur, except for flashes of light as we dropped past side shafts leading to different seams. Air whipped us about in a vertical tornado. I looked up. Above, a square patch of daylight was disappearing rapidly, becoming only a small point of grey light. I hoped fervently the engineer at the controls wasn't sitting there reading over his union contract or catching shut-eye.

Then the blurred walls slowed, and I gained weight tremendously. The coal seam on the level where we got off was about seven feet thick—not much, compared with western coal, but this was anthracite, highly compressed. Mr. Ceeling said we were down about a thousand feet. The air down here was chilly, but nothing like the zero weather above. As we followed Ceeling down a passage, it grew warmer. "It's not bad down here, winter or summer," Ceeling remarked.

We crossed a couple of entries and came into a small room. The foreman flashed his light up onto the ceiling. A row of protuberances ran across the ceiling, spaced evenly along the grey sandstone. It was obvious the marks had been preserved in the same manner as all tracks and marks preserved on the roofs of all mines. Imprints in a soft layer of peat or swamp muck had filled in with sandy mud, and were part of the hard ceiling. Removal of the coal revealed these marks. Each mark showed five sharp gouges in front of what appeared to be a broad, round heel. The prints were all in a single line, all made by the same object, all quite identical.

"What do you think of them?" Davenport asked.

I didn't know what to think. The trail was not that of any quadruped. Or biped, had bipeds existed in Carboniferous times. It was simply the same print, repeated. It could not possibly have been done by any animal.

"It looks," I said, "like a series of marks left by a five-pronged root, on a rolling log being swept along in shallow water."

"A log?"

There was plenty of plant life in Carboniferous times, but it was much different from what came after. Logs? Well, not logs as we know them. But some of the great tree-ferns or giant calamites, forerunners of the cycads and conifers. A huge calamite stalk, uprooted and rolled along, might have made this repeated print with its torn-up roots.

"Look," I said. "When this stuff was laid down, feet were just barely being invented. This couldn't be a creature's footprint; they're all identical. Creatures then had enough to do just learning to use feet; I can't imagine one of them hopping along on one foot, holding the other one in his mouth."

The "oldest footprint in the world" died a hard and lingering death, particularly since I couldn't offer a solid substitute. An "imported scientist" from the great American Museum of Natural History, from New York no less, had let Wilkes-Barre down. The place cooled off a bit. Ceeling went about his normal foreman's labors. Mr. Davenport found he had another engagement. Ceeling helped me with lights to shoot pictures of the tracks before he left. One of the miners stopped while I was gathering up my camera equipment. He said there were some funny marks on the ceiling off a piece in the mine; would I like a look? I would indeed. There was, in fact, nothing wrong with amphibians learning footwork with their newly-created feet here in a coal swamp, no more than there was with mystery dinosaurs looking for salads in a far-away swamp in a far-away time. I followed the miner. He led to another miner. They knew other miners who had noticed funny-looking marks on the roof.

An odd slab of loosely hanging slag near an entry was first brought to my attention. Scarcely three feet wide and not much longer, with a row of thin, regularly placed marks running across it. I found a box at the side of the room strong enough to stand on, and got a closer view. The marks were thin and faint but definitely footprints. I turned my light to catch them in higher relief. The complete outline of a long foot, the heel broad and flat, the toes widely splayed and stumpy. Immediately in front, a similar but smaller foot, a forefoot.

I was led to other funny-looking marks, which also proved to be footprints. Thin, but real. At one point I found a wiggly impression between tracks, probably the light touch of a dragging tail.

Ceeling, the foreman, passing by, looked in on me. "What have you here?" he asked.

"We've found, or been shown by your men here, what you fellows thought you had back in that first room: some of the oldest footprints in the world."

He looked over the prints I pointed out. "Why, I never even saw those things before!" he exclaimed.

"Well, they are fairly faint. And hard to see, up here in the corner. Any way we can pry off this thin section?"

"Easy. We can do it in half a minute with a crowbar."

A crowbar soon appeared, and Ceeling carefully poked and prodded it into the bedding plane. The block came free easily. I broke away excess rock, and we carried it out to the shaft.

In the afternoon Mr. Davenport came back to take me on a tour of other mines. I hoped, if possible, to locate a prospect suitable for display with the tracks of our mystery dinosaur, sort of a before and after view of footprints on the ceiling. I was surprised, as was Davenport, to find tracked-up ceilings were not actually in short supply, once the miners learned what we were looking for. These funny marks on the ceiling were widely known, but nobody had paid attention to them. They were, of course, much more lightly impressed than the dinosaur tracks I had seen on mine roofs. Not only did dinosaurs for the most part outweigh amphibians; the little fellows of the earlier period were hardly yet used to feet, more uncertain about what to do with them.

Once management recovered from the initial setback I had given them by repudiation of their own chosen oldest footprint in the world, I regained status for the museum just a bit. I carried the search into the next day. In a little-used mine I was shown odd marks on the ceiling which I was able to officially bless in the museum's name as assuredly among the oldest tracks in the world. Three led over to a point piled high with slag and stone. Mounting this, I was able to get my face up close to the ceiling and to make certain they were truly tracks, sequential, slender-toed, splayed like baby fingers, with clear heel marks. These were fairly deep, possibly made by an overweight creature. A second trail crossed this one at right angles. A third appeared a bit to the right, long toes projecting, slightly pigeon-toed, showing a knobby heel. This last was clearly a different type of creature. An amphibian playground? It is hard for us to view either amphibians or reptiles as engaged in play. Perhaps it was only a busy crossroad.

Fortunately for the cause of public relations and publicity, I had salvaged a good bit of the mining company's self-esteem. But the state senator had long since left, the local reporter had run off on a dozen stories, and the story lacked impact by the time I had dug up facts to fit. Even to me, the tale had run down a bit when I got back to my hotel. Awaiting me was a telegram in a yellow envelope: "Arrangements made for Texas. Return New York at earliest. B. Brown."

3 1

BACK IN BROWN'S OFFICE I listened to my chief outline plans. All work in Wilkes-Barre was laid aside for the time, worthwhile though it was. I was to leave for Texas immediately, but not for Glen Rose.

Brown told me, "E. H. Sellards, of the Texas state department, tells me two young geologists just found a new sauropod trail down in Bandera County, near San Antonio. They were checking out the story of a couple of local kids who thought they had found elephant tracks. The conditions may be more favorable for removal than the Glen Rose tracks."

The abrupt change in plans didn't matter to me; the basic story was on the front burner now anyway. The Glen Rose story, which had lain fallow and unplowed a half-century, had now been widely reported in *Time* and *Natural History*. This shot of publicity had, of course, set people to hunting tracks of the big flat-foots in a way they had never been sought before; it was natural they would be found elsewhere than at Glen Rose. Time would undoubtedly turn up sauropod tracks in other places.

"This new site," Brown went on, "is in the bed of Hondo Creek, a little stream practically dry most of the year. The track-bearing ledge, which also carries carnosaur tracks, is high and nearly always above water."

Dr. Sellards himself had not seen the new track find, but this seemed of no consequence. If in the end the Glen Rose site seemed for any reason preferable as a source of tracks for the museum, we could go either way. I was anxious to see the new find.

"One way or another," Barnum said, "for one place or another, we'll want a complete double stride. We'll put it on the base of the brontosaur mount, under the tail."

We went downstairs to measure the space. The idea, I thought, was a stroke of genius on Barnum's part. The Texas sauropods, in any case, were not *Brontosaurus* itself. But they were the same basic type of animal, roughly the same build. And they had made their footprints in the sands and mud of early Cretaceous time with meat-bearing feet; to the viewer, the skeleton feet of Barnum's brontosaur should come close to seeming the right size. It wouldn't do to have the brontosaur look as if his feet were pinched in his tracks, but we felt this unlikely. This, of course, would be a matter of concern only after we got the tracks.

The yardstick told us we had a space twenty-nine feet long and eight feet wide between the hind feet of Barnum's giant and the end of the platform. With the contemplated slab in place, it would appear the mounted skeleton, as a living creature, had just walked out of his footprints.

A week later I parked the Buick in front of the office of Dr. E. H. Sellards in Austin, Texas. The man with whom Brown had negotiated the deal betwen the museum and the Texas state department was Sellards's assistant, Glen Evans. Sellards told me, "Evans has made all the arrangements. Ten laborers will report for work tomorrow at the Bandera courthouse. Downstairs, you'll find all the tools you'll need."

I felt as Brown must have felt when the Union Pacific put a steam shovel at his disposal in an earlier day: the world was mine, all mine, and there was nothing I couldn't do. As for cutting out two simple blocks—nothing to it; a simple excavation problem. Of course, I hadn't seen the Bandera tracks yet, but that little detail was a mere matter of time. Off to Bandera!

It was only a few hours between Austin and Bandera. The area was very much like Glen Rose, still a part of the Texas hill country, the same cedars clothing the same sort of slopes found about Jim Ryals's farm outside Glen Rose. The same limestones here represented the same offshore muds of long ago; these low Texas hills had been beach on an expanded Gulf of Mexico at the same time as the Glen Rose area.

When I looked out the hotel window next morning, snow was falling. For Bandera County in southern Texas, snow was as much a novelty as it is in Mexico City. Inside the dreary corridor of the little courthouse, I found waiting for me not ten but one workman. Julius Marquis was a middle-aged man with a shy, easy smile. I told my helper it was good to see a cheerful face in the crew, and he grinned.

"Folks tell me I got a head like a tack, pointed end up," he said. "But I figgered even if we was to do nothing today, I could learn anyhow what this job is about."

The snow, turning to rain, drove another workman in out of the cold. Fred Berg, big and friendly and likable, was a bit younger than Marquis. By nine-thirty the rain had stopped, but no more of my crew appeared.

A small map prepared by Evans showed the location of the Hondo Creek tracks. The site was in the southern part of the county, near Tarpley. I turned to my companions; I was eager to see what the place looked like. "Would you like to ride along to Tarpley and look this thing over? I'd sort of like to see what this is all about myself."

Fred Berg spoke up, "When I first heard about this dinosaur job, I thought somebody was crazy. But if I could see what we're after, it might make sense all around."

So we headed out for Tarpley. The snow, of course, melted almost as fast as it had fallen. Despite the snow and rain, the bed of Hondo Creek was barely more than wet. The road ran along the bank, and a farm house indicated on the map showed up on our right. The spot marked "X" on the map indicated the tracks were near here.

There were carnivore tracks out near the center of what is today Hondo Creek. The tracks ran down the middle of a stone apron, and even riding by in the Buick, I could see they were a respectable size and that they had been made in what was a firm mud long ago. Other trails, many worn and often scarcely distinguishable as tracks, ran here and there. I drove the Buick down onto the firm rock bed, and we all got out.

The carnivore trails led into the area marked on the map. Despite the dirt covering much of the surface, it was easy to see the convergence of these trails. A series of wide depressions showed where one of the creatures had stepped into the other's footprints. Some of the resulting marks resembled large five-toed tracks. But where were there any brontosaur-type tracks? I crossed the area and back again, looking over the superimposed carnivore footprints. There was nothing else here!

This was a heartbreaker. Apparently the sauropod trail reported here had never existed. I could see how anyone unfamiliar with sauropod tracks could have been misled by the double carnivore trail, particularly where tracks had been superimposed. But no one could tell me how to get off the hook: here I was with a crew of ten laborers engaged for the job and nothing to collect.

A man came out from the direction of the farmhouse just across the creek, jumped the thin flow of water, and joined the group.

"Are you looking for what the boys found here a couple of weeks ago?" he asked.

I still hoped I might have misread the map. "Tell me," I asked, "is this the spot where they were supposed to have found a new kind of tracks?"

"That's right; them's them. What do you think of them?"

"Well-l-l, these are not the kind of tracks I'm looking for," I replied. "The boys mistook them for something very different."

"You mean there's no job for us after all?" Fred Berg asked.

I did some thinking, combined with a bit of gambling. Any chance that prospecting might produce another discovery like the Glen Rose find, here, two hundred miles away? Well, the underlying formation was the same as that at Glen Rose. The exposures were much the same. Certainly the country had never been worked over by anyone with any sort of background in tracking down dinosaurs. I explained the situation to my two men, and we returned to Bandera.

Next morning two to three inches of fresh snow lay on the ground; even the weather seemed to have conspired against us. I asked Fred to tell the rest of the men I would get in touch if and as soon as I had something to work on. If things didn't improve soon, I'd have to tell Sellards there was no way to use the money allocated efficiently.

I drove north of Bandera to see if I could find a creek bed Berg had said contained open exposures. A likely spot adjoined a strip of pasture land. The snow stopped, and if I waited, it would melt quickly. But waiting in the car for this to happen seemed intolerable. The ground was white everywhere, but I thought depressions and open faces on ledges would show through the snow, even as tracks on the Paluxy had shown through mud. I set out for the creek.

For an hour I sloshed along in the cold, wet snow. My feet grew numb as I pushed snow out of one depression after another, only to find them nothing but holes. I exhausted all the possibilities of the mile-long exposure. Disgusted, I went back to Bandera for lunch. While eating I recalled a bit of advice Berg had given me. "You ought to talk to Mr. Hunter. He knows the rocks in this country better'n anybody."

I found E. Marvin Hunter in his office. He published a small magazine, *Frontier Times*, and operated a small historical museum. Though he was in the midst of a deadline, he stopped to give careful consideration to my plight.

"Just maybe I can help you," he said. "A year or so ago some girls came to me, saying they had found elephant tracks on the Mayan Ranch near here. I took a look at them but don't know what to make of them. They're about two feet and better across and almost round."

My thought flashed back to the round tracks on the Purgatory River near Pueblo, Colorado. Were these more of the same round prints I had seen there? Mr. Hunter volunteered to go with me to show me where they were. We went south in the Buick across the Medina River and were soon rolling along a road leading toward a few houses and a corral. We came to a fork in the trail leading to a bend in the river, and Hunter told me to follow it. It came to an end below a bluff. Hunter got out and I followed him along a flat exposure. The snow had melted, and the rocks were drying off.

We came upon a series of round depressions, much as if a washtub had been pushed into the mud before it had turned to limestone, at staggered intervals of about eight feet. My hopes rose. I went to work with my whisk broom and brushed some of them out. The matrix had been ideal for receiving the imprint; most of them were as sharp and clear as the day they had been made. The passage of well over a hundred million years had not marred the fine detail. Characteristics I hadn't been able to see in the Purgatory prints were clearly visible. There was no mistaking the outline of the great pad of flesh that covered the toeless foot. The prints had been made by the left and right forefeet of a sauropod dinosaur. I walked back and forth, trying to make sense out of a trail I could see clearly but could not read at all.

"Do you have any idea what they are?" Hunter asked.

"Know what they are," I laughed. "They're the very thing I came to find. But they're not all here."

Hunter looked a question at me.

"No hind feet. The prints of the back feet aren't here!"

I felt a bit sillier than the dinosaur must have looked. A forty-ton creature walking along on his front feet, his tremendous rump balanced in the air?

One end of the trail had been worn away by the water, but the other end ran directly under the nearby bluff. If only I could see more tracks, they must tell me . . . Then the answer struck me. I had a whole crew of men just waiting for something to do!

"Where's the owner of this ranch?" I asked Mr. Hunter.

Half an hour later we were headed back for Bandera. I had met the ranch foreman and had permission to uncover more trail. I took Mr. Hunter back to his office.

Next morning I had a full crew at work tearing into the heart of the bluff. They might enlarge the question or find the answer. I still needed footprints to fit behind Barnum's brontosaur. There wasn't room in Dinosaur Hall to set up the great beast upon his front feet. So I left the job in charge of the best man I had as foreman and went on prospecting, taking Fred Berg with me as scout and guide.

We were off first for the nearby Davenport Ranch, because Hunter told us there were some extremely good carnivore trails there. I certainly was in no need of carnivore trails, but where there are trails in plenty there may be some we want. So we went looking. In only a short while we got to the ranch house, where a young man opened the door. I asked about Mrs. Davenport, owner of the ranch.

"Mother's not here," he told me. "She's visiting friends in Uvalde."

"I'm looking for dinosaur tracks," I said. "Mr. Hunter in Bandera told me you have some interesting ones here."

"Well, we've got quite a few of them down on the creek."

Fred spoke up. "Mind showin' 'em to us, Smokey?"

Minutes later I was looking down on the finest

carnivore tracks I had ever seen. The single trail ran under the bank and could be traced for ten clean-cut steps.

"Mother's had her a time of it with these tracks," young Davenport told us. "Somebody's always coming along wanting to cut prints out for souvenirs. I'm sure she won't mind you looking at them, but I know she won't let you have any."

I walked along the bank, admiring the fine trail. It led to an area covered with potholes: these were clearly not the tracks of other carnivores. My companions watched as I pushed out the leaves and rubbish from one of them with my whisk broom.

There was a large, broad, lizard-like footprint. Four great claw marks came to light under the broom. I looked up at Smokey Davenport.

"Did your mother know these were here?"

His expression told me. "What do you mean?"

"I mean these odd, partly buried depressions."

"Are those tracks, too?"

I uncovered a few more. There seemed to be four or five trails leading directly under the bank, along a twenty-five foot exposure. I stood up. "Is there a telephone where your mother's staying in Uvalde?"

Trackway of the floating sauropod, Mayan Ranch, Bandera, Texas. Photo by Bird.

MRS. DAVENPORT proved to be as big a problem as any in the field. She was pleasant enough over the phone. She was most open to permitting us to view, even to clean the dirt out of any tracks on any of her property. She was obdurate about any action beyond this. No one had ever removed any tracks from Davenport property to her knowledge, and not even the prestigious American Museum of Natural History of New York was about to do so now or at any future date. Any charm exuded by the said museum's field representatives was as so much sweetness and fevered breath wasted on the desert air. Gathering any charisma I may have possessed in those days, I managed to wring a promise from her, with a promise I was wasting my time. She promised she would wait up and tell it all to me to my face. So off to Uvalde.

I crawled into bed back at the hotel in the very last of the wee hours, tired in body, with ragged tonsils, sagging sadly in both vocal cords. Magnificent in defeat, but not quite routed. Truly Barnum Brown was not the only man who had put forth tremendous effort to secure aid and cooperation on this project. Mrs. Davenport had long been intrigued by the tracks of the carnivores that had once trespassed on her properties, so I had something of a foot in the door. She listened to my tales of the far off mystery dinosaur, with photographs which I just happened to have with me. I showed pictures of the Pennsylvania coal mine amphibians, from a hundred million years before her dinosaurs. I talked learnedly of the contribution Davenport Ranch might make to science. In the end, she agreed to let us move in with picks and shovels whenever ready. Dig all we want; just don't take anything.

All this was, at the time, a great promotion job. I knew what we wanted to find; I just didn't know where it was nor if it was. I still had hanging over my head a mind-boggling load: forty tons of dinosaur stumping about back at the Mayan Ranch, tromping down the mud on his great front feet, and what did he do with the balance of his great body? And how did he balance it? And why?

Fred Berg and I went back to see what had been done by my crew. The crew had cut twenty feet or more into the footwall of the steep bank. I was astounded by the mounds of rock and earth that had been moved. And when they started to clean up the operation to see what they had found, I was astounded more.

"Hey!" Bud Kalka, one of the gang, exclaimed. "The trail's making a turn!"

I stepped into the created cavity. Bud was right; the great beast had turned, and sharply. I followed the tracks as far as they were exposed. Only one differed from the rest, at which point the trail made a turn of about thirty degrees. The different print, dug in more deeply, made the story of the whole trail clear. It was only a partial footprint, but it was a broad rear foot, with claws showing clearly, just the front part of the foot. The big fellow had been peacefully dog-paddling along, with his great body afloat, kicking himself forward by walking on the bottom here in the shallows with his front feet. Deciding to turn back to deep water or onto land, he had kicked his body into an S turn with one push of a hind foot! This was a great relief to all of us reading sign.

Up to this very important point, however, we had found no tracks to fit our basic goal: a set of prints to fit Barnum's brontosaur back in New York. So we packed up and moved, not wishing to give Mrs. Davenport too much time to think things over.

It was a cold morning when we got there.

Little needles of ice had crept in the night out from the edges of the puddles, but when the sun came up, this retreated. The men swarmed from the cars down the fifteen-foot bank to the track-bearing apron of stone below. I set five to work uncovering more of the carnivore trail, while the rest of us went to work where the sauropod tracks disappeared under the bank.

Work went along rapidly. By the end of a week we had cut out a great patch of the overlying limestone, leaving a protective cushion of soft shaly clay over the maze of welts and bumps and potholes that marked the edges of new tracks. We established the lateral extent of the trackway and turned to concentrate on the space over which the sauropods had walked. The quarry was now ninety feet long, five to ten feet wide in places. Like the Howe Quarry, the more we found, the more there was left to be found. As Bud said, "Looks like we could go on uncoverin' these things for months."

Mrs. Davenport returned from Uvalde presently, and the red-headed widow was a frequent visitor at the quarry, becoming as interested in the tracks she had never suspected were there as she had been in the long-known tracks. My own interests in the area were in no wise abated by the fact there seemed to be no end to the number nor to the length of the sauropod parade. They continued to march stolidly into and under the fifteen-foot embankment. That these might have been the same unnumbered horde who had gone to their grave in Howe Quarry didn't seem quite possible; not only was the trip too far, but they weren't headed in quite the right direction. Nor were the Davenport

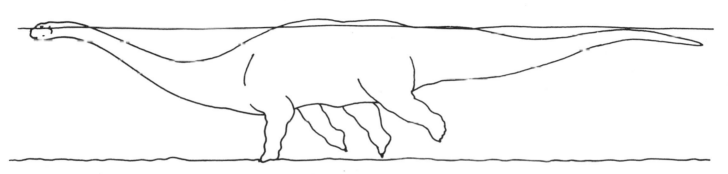

How the Mayan Ranch Trackway was made. Copied from a 1944 drawing by Bird.

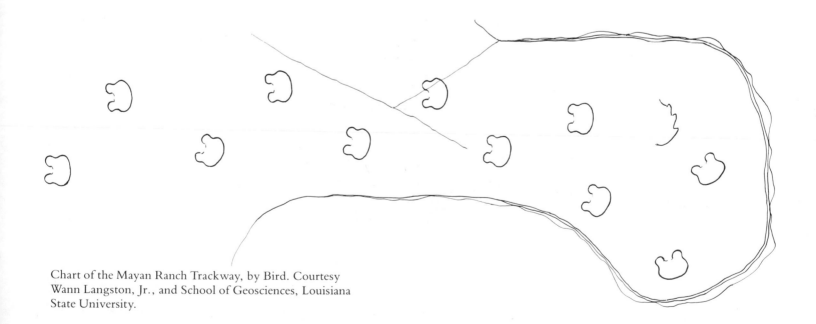

Chart of the Mayan Ranch Trackway, by Bird. Courtesy Wann Langston, Jr., and School of Geosciences, Louisiana State University.

Ranch and the Howe Quarry sauropods of comparable age.

After a week more, I couldn't wait longer to see exactly what we had come up with. I set the crew to cleaning off the shaly overburden we had left to protect the tracks while wheeling away the material we had removed. We marked off a fifteen-foot wide path for cleaning right down to the bare tracks. The soft protective layer was, of course, no problem at all. But unfortunately the limestone layer in which the tracks were incised was a soft limestone. When I set the crew to work with whisk brooms and awls instead of shovels and wheelbarrows, there was some question in their minds; they were afraid their production would appear quite puny. Adolf Kruckermyer, the foreman, said, "We'll be to the end of February doin' this housekeepin' stuff." He wasn't wrong. In a few days, though, when they got used to the work and got over the idea I would judge their production in terms of wheelbarrow-loads, they settled down to become seasoned museum hands instead of pick-and-shovel people. More important, they regarded themselves as servants of science.

In this they were fully justified. Before they took on this job, everything they had revealed here had lain ready to be seen for thousands of years, even before Man was here to see it. Never had Man looked on a sight like this. Here was not a single sauropod trail as I had found on the Paluxy; here a herd of giants had stampeded, or moved, as a single entity. I tried for an accurate count, but it couldn't be done. There were to begin with prints of seven individuals, in a twenty-foot space, but a few feet beyond this well-trodden area the tracks broke down into a hodge-podge. One fact stood out above all: they moved in the same direction, and presumably at the same time. This was not true of the carnivore trails; there was some evidence, in the amount of erosion in various tracks, that these had been more random. But the great flat-footed giants had been a body. Moreover, there was evidence of rudiments of a mammalian-type herd instinct; the tracks of the little fellows were in the inner part of the pathway.

I looked in vain, however, for a single suitable trail to fit behind Barnum's brontosaur. The only animal that walked alone had feet too small to fill the bill.

I wrote Brown about the situation, telling him

Working at the Davenport Ranch site.

that subject to approval I would enlarge the excavation in hopes of finding divergent trails that might suit the purpose. Brown wrote back promptly, approving what had been done and leaving the next moves up to me. Sellards came around in person and did the same, expressing a wish to have a sauropod track for the university but leaving its source up to me as well.

An amusing problem in science came up with one of the prints, the largest sauropod track. Bud Kalka came up with an oddly shaped lump of rock, fished out of the very largest print. It was in the bottom, a loose and oddly shaped limestone lump. The track from which it was taken was oversized even for a sauropod hind foot, oversized obviously because the spot had been stepped in by chance repeatedly. It must, naturally, have been a good bit on the squishy side; the last fat flatfoot out of that depression dropped a sizable gob of mud either from the foot or knocked from the rim of the print, and this lump had petrified, like the surrounding track. There are thousands of dinosaur footprints scattered all over the world, but identifiable gobs of mud from a dinosaur's foot are exceedingly scarce.

The longer we worked, the deeper we dug into the bank, the more confusing the trail became. I resorted finally, to have a clear record of the project, to the same technique I had used at Howe Quarry: I blocked the finished area off in squares with strings, and went to work on a detailed quarry chart. Such a work is not too rare in this field; I had in fact made such charts for single trails. But because of the intricate herd movement, I suspect few charts of such complexity have been made.

As a Davenport Ranch finale, I located two new sauropod trails, a bit removed from the rest.

The creatures had been of medium size, perhaps forty feet long, give or take a bit. They were, I knew, not suitable, because they were somewhat eroded. But with them was one furrow, a bit better than four inches wide by four inches deep. I asked Julius Marquis to clean it out, to see what we had. It ran squarely between a set of prints. A few minutes later, I looked over Julius's shoulder. The furrow was not a waterworn trench; it was smooth and uniform in the bottom as along the sides.

"What have we here?" I asked.

"Looks like one of those big fellows was draggin' his tail."

The chart brought back to mind questions about the habits of these great sauropods. How deep had been the water through which this herd had moved? There were no tail drag marks in what had been the limey mud they had waddled through. There was no indication the little fellows had been swimming at all; their prints were as sharp and clear as the rest. Even had they come up well on the little fellows' flanks, they would still have rippled below the belly of their large companions. The adults must of necessity, then, have carried most of their huge bulk out of the water. Yet no tails were dragging.

What about this big fellow? Is the last dinosaur in the parade supposed to sign off with his tail? Was he more tired than the others? Had he just outrun a meat-eater?

At this point, we don't know. The tail drag marks in dinosaur prints can be counted on one hand. Perhaps scientists of future ages may be able to settle this. Meanwhile, most of my dinosaurs run high-tailed.

Drawing of the Davenport Ranch sauropod herd.
Copied from a drawing by Bird.

33

APRIL 7, in Glen Rose.

All efforts to locate suitable sauropod trails on West Verde Creek having proven futile, Dr. Sellards had made arrangements to have a quarry opened in the bed of the Paluxy.

I drove down to the courthouse square before eight the next morning. A new crew of men was gathered by the little white limestone building. Some sat in the shade on benches, others stood about in the parking area.

I recognized John Mathews from before, among the standing group. He was old, grizzled, nearly toothless, but there was a strength of character and a toughness about his wiry frame and a twinkle in his eyes. If I needed a foreman at times when I might be away from the job, John would be my man. In tattered, faded overalls and wearing a slouch hat that had seen the very last of its best days, he fit in well with the rest of the group. A good number came from Cleburne, nearly forty miles away. They were drawn by promise of a steady job while it lasted, plenty to drink—we'd be working in the bed of the Paluxy River—and good pay: a dollar a day. Ones with special skills, like with rock drill or sledge and chisel, might rise to $1.25.

John recognized the old Buick as I came around the corner. He waved and gathered the group together. Manuel Gosset and Monroe Eaton, young men in their twenties, were with him, as well as a tubby character from Bono, a cross-roads settlement on the way down. There was Tommy Pendley, a young family man in his twenties, and Oscar Moxon, and John's brother, Sam.

"Well, boss, we're all here," Mathews greeted me. "Leastwise, all but Harry Shoemaker, and he'll be here if he kin whup his old car this far. Where's them dinosour tracks? We're rarin' to go!"

We piled into the cars, and as I swung the old Buick away from the curb a four-wheeled clatter bore down on us. It was Harry Shoemaker in his Model T.

"You'd better git that old rattletrap into the parade," Mathews shouted. "We waited fer you this time, but we wasn't gonna wait much more."

The river was up at the first ford, higher than I'd ever seen it. The old Buick swam through fearlessly; she had done this all before. In the rear view mirror I watched the cavalcade behind me dunk in and pull out, streaming water from their running boards. By the time the last car had drained its floorboards, the first car had nosed into the second crossing. And in another minute or so we headed down the live oak lane on the bluff at Jim Ryals's place. He looked up in surprise at three dripping cars loaded with manpower, picks, shovels and paraphernalia.

"I told you I'd be back one of these days," I called out. "This is the first one of them."

We parked the cars on the bluff before the fourth crossing. In the back of the Buick were the tools we had used on Davenport Ranch, plus around a hundred and fifty burlap bags.

"We'll need all of those bags and three or four shovels," I said. The prints I had come upon on that first day were of course mud-filled again. I wished that I could have had at hand then the resources I had now. I explained to the crew a little something about dinosaurs and dinosaur tracks. Most of them knew the three-toed tracks for what they were; probably a few of them had chosen track-removal in preference to cedar post-cutting and moonshining a time or two in their pasts. But this was the first spot in the world where the big sauropod tracks were recognized for what they were.

"To start off, we'll have to build a coffer dam," I told the group, "so we can shut out the water.

After we get a look at what we have here, we'll know what to do next. First, let's get rid of a lot of this mud and sand by stuffing it in the bags." Actually, there was lots of good sand on the bluff above the ledge.

I produced a sacking needle and a ball of twine and gave them to Moxon. "You look like a good man to sew them shut," I told the stocky little fellow. Then I showed Monroe Eaton and Tom Pendley where to begin to run the dike. It would be necessary for them to work knee-deep in water, but the water was fairly pleasant, not at all really cold.

I went to work shooting pictures of the procedure. Brown would want this better covered than Kitty Hawk or the landing of the Pilgrims. This was an important first. Of course these sauropod potholes had been known for years. But, like the "big bird" tracks, they had never been known for what they were. Natives of the region who had noticed them at all had regarded them as curious potholes put there by a bountiful Mother Nature to make it easy for boys to catch stranded fish by hand when the river was too low for sport fishing.

As I turned to go for the movie camera to shoot the first pictures, there was a great splash behind me. Monroe Eaton, toting a bag of sand, had stumbled into one of the sauropod prints. Between Monroe and the bag of sand, they made a fine thing of it.

Streaming water from every point, Monroe came back up, with a question. "Mr. Bird, did you say those big guys were called thunder lizards because they shook the ground when they walked?"

"That's what they say."

"Well, I don't know nothin' about shakin' the ground walkin'. But standing still here, the son-of-a-gun shook hell out of me!"

The coffer dam began to take on length. It stretched downstream from the discovery ledge like a long brown finger pointing toward the spot where the water-covered trail ran under the bank. The solid wall of sandbags rose about a foot above present water level. Noon came, and I discovered the burlap sacks were running low; it had taken more than anticipated. I spent the early afternoon visiting farms along the river, cadging a fresh supply. Mrs. Ryals found thirty-five in her feed room. One farmer sold me sixty; another, eighty.

Just after dawn next morning John Mathews, with his whole family on a platform trailer loaded with canvas, passed the Lane Hotel while I was still at breakfast. When I got out to the fourth crossing, the Cleburne men had set up camp. Eaton, Pendley and the Mathews brothers had brought their wives and children and equipment enough to live out along the river. The weather and the site were ideal, the quarry only a ten-minute walk away. I should have liked to join the party with a cot and a piece of canvas but felt I should maintain quarters in Glen Rose, sort of to act as a liaison with the local people.

We laid stepping stones across the Paluxy just below camp. Yesterday we had jumped from rock to rock at the rapids, but now our stepping stones would allow us all to cross dry-shod. By nightfall the extended finger of the dike lacked only a few feet of completion. Early next day Eaton dragged the last bag into place, and we were ready to start bailing away. The dike enclosed an area nearly two hundred feet long by fifty feet wide. I had brought out ten galvanized pails.

"It looks like a lot of water to bail by hand," I told them, "but that's the way we'll have to move it."

The men rolled their pant legs above their knees, lined up along the dike, and started bailing. I just rolled film.

The water went down a lot faster than I had expected. This operation was different from uncovering fossils with a shale or sandstone overburden. We were about to lay bare several thousand feet of surface in a few hours. On the face of the sloping discovery ledge, the holes I had felt with my feet that day a year and a half ago now showed up as cavities under mud. I took a shovel and started to clean out the first. Mr. McCoy, the man from Bono, worked close by. I called him over.

"If you'd like to see what we're after, get a broom and lend me a hand."

The old man joined me, cherubic in bare feet and with overalls rolled above his fat knees. He had become known as "Tubby" and, as a workman, I was well pleased with him. He helped me sweep out one of the big depressions and stared in astonishment at the enormous outline. "Well, I be dogged if it ain't a track, sure enough!" he said.

Monk Eaton and Tom Pendley brought buckets of water and splashed it into the broomed footprint, washing away the last traces of dirt. Moxon and Gosset came over to look. John Mathews came but returned to urge the crew into a steady bailing that dropped the water level behind the dike an inch every few minutes. The project had taken on new and clearer meaning to everyone.

The swish and splash of water became a rhythmic music. A few small leaks were noted, but Tom and Monk soon stopped them by stuffing empty bags into the right spots. The river swirled behind the dam like the top of a shaking table.

The boys finished the bailing before lunch. It took but a few minutes with all hands shoveling and washing the exposed floor to clean up the hard, flat bottom encased by our dike. I counted forty-two prints to the point where the trail disappeared under the low outcropping ledge at the south end. All the tracks were somewhat water-worn but in fair condition on the average. The great trail might have been made by Barnum's own brontosaur but for the fact he was found over a thousand miles away and lived millions of years earlier.

We ate lunch up at the campground. The four families had quickly adapted themselves to life under canvas. The men were, of course, far outnumbered by their women and children. I got an impression the majority bore the name of Mathews or were related to people of that name. Laughing, happy kids were everywhere. Mrs. Pendley gave me a bowl of hot stew to go with my hotel sandwiches. As their menfolks finished eating, the women gathered up the dishes, and talk turned to what we had uncovered in the morning hours.

"Where do you reckon that animal was headed?" Mathews asked. "Looks like he had some point in mind, the trail's so long and straight."

"Well, that's got me guessing, too. We'll try to find out about that this afternoon," I said, as I rose to return to our job.

The exposed tracks were surprisingly good, considering the erratic actions of the Paluxy. But before I undertook the task of cutting slabs from the river bed, I wanted to look into conditions under the bank ahead. In addition to the big trail, we found one of a smaller individual. It came in from under the dike and paralleled Trail One. We carried crowbars, sledges, heavy drills and chisels down to the spot where both trails disappeared. A ledge of limerock some fifteen inches thick lay over the spot, fractured into masses about ten feet square.

"Now boys," I said, "we want, next, to move

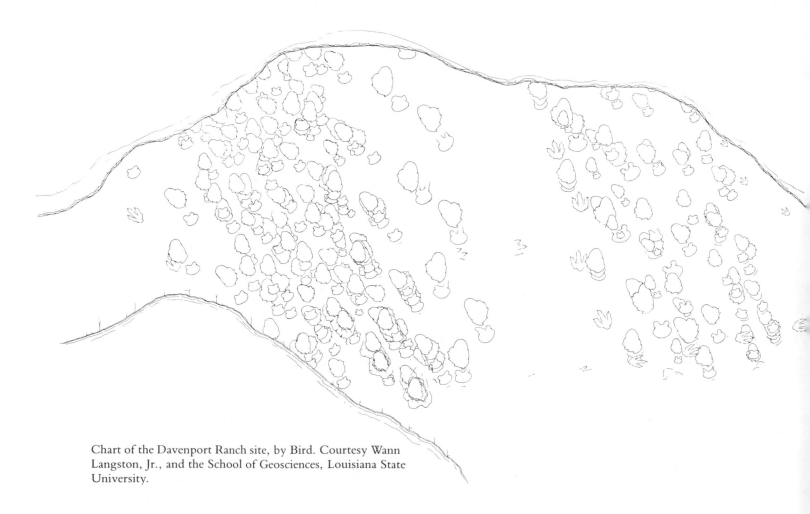

Chart of the Davenport Ranch site, by Bird. Courtesy Wann Langston, Jr., and the School of Geosciences, Louisiana State University.

Sauropod tracks at the Davenport Ranch site.

some of this ledge, so we can follow the big trail. It'll be easy to bar the rock out of the way into the river."

Monroe Eaton and most of the other men picked up tools and began looking for favorable places to start, but McCoy stared at me as if he hadn't heard clearly.

"Why, now, see here, Mr. Bird," he blustered. "You're kidding. You know that animal never walked under that there rock."

I looked at the man from Bono. He wasn't fooling; he was genuinely incredulous. Because there wasn't room for everyone on the ledge, I set four of the men, including McCoy, to moving a heavy layer of drifted sand and clay from the west side of the cleared area. There were no tracks there, so far as I knew, but it was a site that offered easy exploration. McCoy worked with this group, and his round face was a sight to see when they came upon another track.

Half a dozen women and children came down from the camp and stood on the bank. They looked curiously at the dry floor of the dammed-off area and its exposed trails.

"Well, y'all sure enough got something!" one of them exclaimed. "I hadn't took much stock in this 'til now, but this just goes to show! Me, I'm a-goin' back to camp, afore I take a notion to help dig. It does beat all, what's laid here in the river all these years, and none of us a-knowin' it."

The men sweated away at their respective tasks. Monroe Eaton and Tom Pendley worked slowly but deftly, sliding a great mass of rock off into the river, prying it along with their bars. Close by, Manuel Gosset swung a ten-pound sledge against a great rock mass which he had pried up onto its end so his blows might break it. Gosset was almost Herculean in stature and muscle, and both maul and rock seemed to be taking punishment. Beyond him, the man from Bono worked, helping John Mathews bar stone. Underneath the limestone stratum, a fourteen-inch layer of soft, green, clay-like shale covered the track-bearing surface.

The men cleared away quite a footage of the thick limerock. Monk and Tom laid aside their bars and began to shovel away the soft clay, digging for the point where the next track should be. The clay was tossed toward the river, where it would keep a rise in the river out of our quarry.

"Take it slow and easy," I warned Monk. "You don't want to scar that track with your shovel, when you reach it."

Out of the corner of my eye I watched McCoy. He turned to watch Eaton, now scooping down into the depression that was the track. The edges rose slightly in a great roll in front of the toe prints. Eaton's shovel rasped the bottom, and I stepped down into the track to clean it out with brush and awl. McCoy watched the great claws take shape, watched me broom out the thirty-eight-inch depression. His jaw dropped below his collar. He sat down on a rock.

"Well," he said. "What d'you know! I still don't see how he ever done it."

He looked about at the surrounding landscape. The river bank rose beyond the ledge to a height of some thirty feet. Down at the bend there was a rise in the hill where two hundred more feet were piled up. McCoy tried to comprehend the significance of tracks following a bedding plane under this mountain of material. "What d'you know!" he said again.

"I think we have something over here, boss," someone called from the west bank.

It was Sam Mathews, shoveling sand. He had encountered another great track. I took my small tools and went to work loosening the last of the clay lodged in its bottom. It was one of the finest of the brontosaur-type tracks I had yet seen, no larger than Trail One, but the creature had walked in firmer mud. The impression was deeper and more clean-cut.

"Boys," I said, "this may be the best trail yet."

The men dropped their tools and came to look. Tom Pendley grabbed a shovel and went to digging where he thought the next track should be. Monk Eaton pitched in to help Tom. McCoy and John Mathews went back to the first big trail. I went for the cameras again.

Work went on. Hours flew by like minutes. Noon came; we stopped for lunch; we went back at it. About two o'clock a group of people, apparently from nearby farms, came down from the fourth crossing. I was hardly aware of them until someone spoke.

"Do you mean to say all those things out there are tracks?"

I looked up and nodded. "That's right."

The group departed but came back in half an hour with several other people. I saw Jim Ryals, his mouth agape, and climbed to the bank beside him.

Bird's Paluxy River crew at work.

The Paluxy River trackway site after a flood halted the removal of the tracks.

Bird clearing off sauropod tracks behind the coffer dam, after water was bailed out, at the Paluxy River site.

He stared down at the dammed-off area, bright and clean in the sun. "That's quite a dam," he said.

Tom Pendley called over. "I've got another trail . . . a carnivore, because he's got three toes with claws. Look, he's walking alongside the big flat-footed guy; wonder what they were talking about."

"Probably lunch. That is, if they were there at the same time." I looked the trails over. They were both sharp, clear, fairly equal. There was no reason to doubt they might have been made at the same time.

Back at the hotel, I ran into Mrs. Lane. "I hear you people have found a whole lot of dinosaur tracks. May we come out to see them?"

What could I say? But I recognized this was the beginning of a tide we couldn't turn off, couldn't offend, couldn't use to speed up our work . . . anything but that. And it was, nevertheless, important. Museums need good will and public interest too. But beyond the Lane Hotel was Glen Rose, Cleburne, and Dallas and Fort Worth only an hour away.

The men had Sunday off, but I went out to the quarry to look things over and lay plans. The first visitors arrived before I got to the planning stage; at five the tide began to slack off. The press, a reporter from *Dallas Morning News*, didn't show up until Monday morning, looking for full details and pictures. Before summer was over, we had put through ten thousand people. Fortunately—and boringly—I needed answers to less than half a dozen questions.

But back to the salt mine: John Mathews's group, working on Trail One, uncovered more of the big flat footprints. The other group, having no rock to hold them back, went like crazy on Trail Three, carrying the paralleling carnivore trail along with it.

"How much more area are you going to uncover?" the reporter asked.

"We don't know. As long as the overburden remains reasonably light, we'll keep on. When we start to take up slabs of specimens, the more material we have uncovered, the wider selection we'll have."

The carnivore trail was a case in point. I called the reporter's attention to it.

"See this three-toed track? A meat-eater."

"Why? How do you know?"

"And that big flat-footed fellow? He won't touch meat. He wouldn't touch you unless you turned green. Now, you see here: where Flat-foot turns, Bird-foot turns too."

"But how do you know they were here at the same time?"

"We don't. We don't even know they were here the same year. But they made the same right turn in the same place. And both tracks are about even as far as quality of the track is concerned."

"So what do you make of the turn? Females?"

"I doubt they were that human. More likely, the meat-eater was trying to urge the big fellow up on solid ground, where they might have lunch together, one inside, one outside. And the big fellow, he was surely trying for deep water; the meat-eaters among the dinosaurs never conquered the waters."

"How do you think it might have turned out?"

"In this case, I don't think. Look at this track . . . and this."

I led the reporter down the walkway two steps beyond the turn. Here was the claw-print of a carnivore overlapping the flat-foot's big print. Following this, the carnivore was apparently hopping along on one foot, the right foot. Did he have a rear claw set in the belly or the rump of the plant-eater at his side? I pointed these things out to the reporter, and I pointed out that the story was headed for the ledge.

"And when you dig in there, you might find an 'after lunch' pile of bones?" the reporter urged.

"Not a chance. First: we don't know who won. Second: the processes that preserve bones and those that preserve footprints are so different that they're never found in the same place. That's what makes it hard to make a case on evidence like this. We just guess our best and hope our guesses will hold up with the other fellows in the same business."

The Sunday following the *Dallas Morning News* coverage brought an avalanche of visitors, fine for public relations but hard on our work schedules. By ten o'clock, on weekends, we could count on seeing the bank above us covered with cars, much of our work area covered with people. Fortunately, our guests were less of a problem here than in Howe Quarry, where our concern was bones. It takes almost a day's work, by an experienced worker, to pick up and walk off with a track made in stone.

Hanging over our heads literally was a much more pressing problem than our visitors, hanging twenty feet high in the trees along the bank. Mr. Wilson, the farmer who owned the property adjoining the bank along the river, brought this

Photograph of the trackways made by a sauropod and a pursuing carnosaur, Paluxy River.

home to us. This quiet man was as astonished as his neighbor, Jim Ryals, at what we were turning up in "his" river, a river whose tantrums he understood better than we.

"You doin' fine," he told us. "But watch out for this here river. If there comes a heavy rain and a good rise, you'll be in a tight."

"How often do you get these rises?" I asked.

"Well, here it is the middle of May. River always goes away down in the summer; it'll almost go dry, sometimes. But summer's still a month or more away. Might be you'll get a big rise before then; might be you won't."

The end of the week came. The limestone and shale had been removed from seventy-five feet of Trail One, and it was heading again for the curving river. Early Saturday afternoon Monk slipped off his shoes and dropped down in water waist deep to feel for the line of tracks. It wasn't there; the monster seemed to have turned even as the stream had turned. Monk slid his exploring feet farther out from the bank and found an unexpected hole. He dropped down into the cavity until only his head and shoulders were above the water. He felt around a bit. He looked up at me, wide-eyed.

"There's a track out here you could nigh about lose a cow in."

I tried to see the depression he stood in, but of

Bird's interpretation of the carnosaur-sauropod chase sequence, lateral view, from a letter to Wann Langston, Jr., August 10, 1976.

Bird's interpretation of the carnosaur-sauropod chase sequence, view from above, from a letter to Wann Langston, Jr., August 10, 1976.

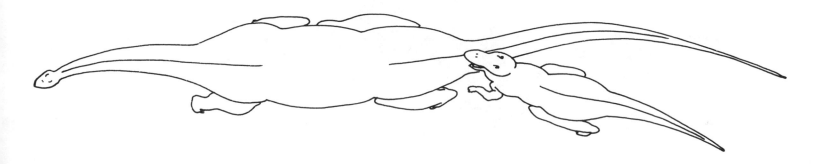

course the water was too roiled by his movements; I had to go by what Monk told me his feet were telling him. He traced the new monster's course. The trail came out from under the pile of dirt and sandbags that was our dike and continued in the same general course as that taken by the creatures of Trails One, Two and Three. Monk stumbled into more potholes farther out. Though waterworn, they were clearly sauropod tracks. Had there been another herd of these long-tailed, long-necked giants passing here, like what we had found at Bandera? A stampede? A beaten path?

"Looks like we'll have to build another dike, soon's we get time, if you want a look at this trail of real big lizards," Monk suggested.

In the early afternoon it clouded up in the west, and I scanned the dark sky anxiously, recalling Wilson's warning, looking up at the trash left high in the trees overhead from previous floods. This foreboding cloud cover always seemed to speed up the crew. Manuel Gosset's hammer and chisel seemed to ring out more loudly in the still air.

A familiar bulk of a man with a shock of dark hair shot through with a touch of grey dropped down from the visitors' gallery into the quarry. My old friend, Ernest "Bull" Adams, the "smartest man in town," the "best lawyer in Texas," was paying us a call in his characteristic manner. He grinned a greeting, but instead of offering palaver he picked up a crowbar and joined the men in heaving on the masses of stone Mathews and Moxon were struggling to dump out of our way and into the river. I owe a lot to Ernie Adams. As unchallenged local pundit his decree that man was firmly entrenched in Cretaceous time, alongside the dinosaurs, had done a lot to protect the Paluxy story from scientific investigation until my time to find it had come.

After a generous contribution of muscle power, Bull laid aside his crowbar and waved a friendly goodbye. "Drop around to see me sometime, when you have time," he invited. "I have some things to show you."

Of this, I had no doubt; Adams was the most interesting and accomplished amateur archaeologist I've ever met.

"I can't do it when I have time," I told him. "I'll have to come before that. Since that thing in the Dallas paper, this is close to being a twenty-four-hour-a-day job. But I'll be around."

I took another look at the cloud bank. It had thinned out and started to go away. It was quitting time, so we left.

Bailing water from behind the dike at the Paluxy River site.

34

SO THE MEN built a second coffer dam, fifty feet long, forty wide, six sandbags high. The volume of water it impounded was considerable. With everyone manning the buckets, the power of the crew to remove water was considerable, too. In an hour the rims of the new track Monk had found began to appear. A little more, and spurts of water squirted up from cavities within the print.

"You're goin' to have yourself a time keepin' this place dry," John Mathews said. "River's comin' in so fast, we'll not be able to stop bailing."

John was right. The prints were so deep that in places they had penetrated the stratum of limey mud in which they were made and extended into the clay base beneath. The great claws had punctured the clay. Here at the edge of the bank, the clay had washed away. Monk tried to stem the flow by caulking the leaks with burlap. His efforts were futile; the rough wadding popped out time and time again. The pressure under the rock was just too great.

"Well, at least we can make a record of what's here," I decided. I stretched measured string to chart the trail, while the men kept the situation under control by bailing. The tracks of the hind feet were colossal. They were considerably larger than any we had uncovered before, but not only because of the creature's size; their maker, in sinking in so deeply, had impressed a great slide with his heel at every step. And foot-high welts of mud were piled up in front of the larger prints. One track, including the disturbed mud, measured two inches under five feet, a respectable size for a bathtub.

The newly diked site filled up, of course, overnight, but I had seen the bottom; I was content. We turned all-out attention to Trail Three; these fine tracks closest to the bank seemed by far most ideal for quarrying slabs. I went about with a fifty-foot tape, going over the prospects. The great slabs, each weighing tons, could not possibly be removed from the riverbed intact; they would have to be broken up and removed piecemeal, as numbered sections.

Two natural features made this practical. The track-bearing stratum, eighteen inches thick, was not solid; it was broken up by cross-fractures at intervals of a couple of feet. Beneath it was an equal thickness of clay. By cutting outlining channels and digging away the clay from below, the slabs could be broken up readily. Some of the tracks we would want still lay under the ledge, and we must knock this down with dynamite. I set the men to work drilling holes.

Coming out of Hilda's Cafe that evening, I noticed dark cloud banks gathered in the west. They looked serious enough to make me hasten back to the hotel. Thunder grumbled and growled a bit, decided it wasn't getting attention, so a wind tore down the hall and started slamming doors and blowing curtains in windows, twisting and tugging at the limbs of a big cottonwood outside in front of the hotel. The rain came down on the heels of a blinding flash. Nobody in Glen Rose was more concerned than I. The river, the river Mr. Wilson had warned me about, the river that piled brush out of reach in the trees above its banks—what would it do? The rain came off and on throughout the night, greatly disturbing sleep. How much would it disturb our work out at the tracks?

Dawn came on sloppy and appropriately dark grey. I dressed, slopped out to the Buick, and wallowed down the road to the first crossing. Crossing? The water was five feet deep . . . five feet deeper, that is. A lot of cars were lined up there,

Bird measuring a sauropod hindfoot track.

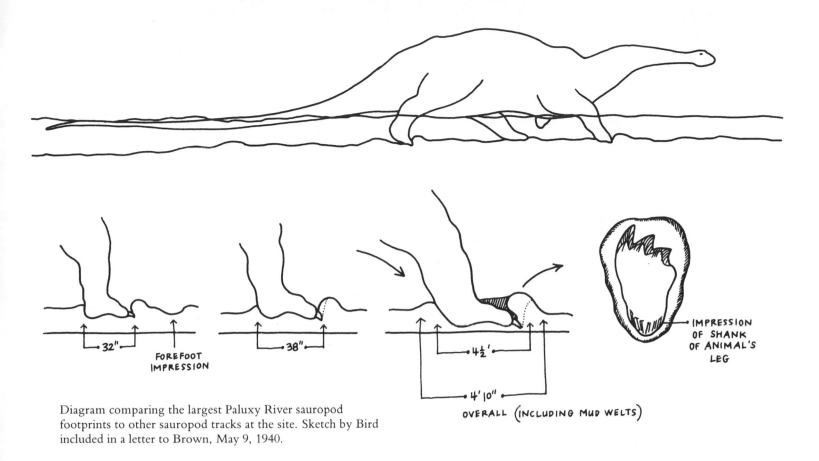

32"   FOREFOOT IMPRESSION

38"

4½'

4' 10"

OVERALL (INCLUDING MUD WELTS)

IMPRESSION OF SHANK OF ANIMAL'S LEG

Diagram comparing the largest Paluxy River sauropod footprints to other sauropod tracks at the site. Sketch by Bird included in a letter to Brown, May 9, 1940.

none trying to ford the river. I walked down to the bank of the still-rising stream. "Might as well forget going out to your diggings today," a man beside me advised. "And tomorrow too."

By ten o'clock on the first of June a few of the seasoned old timers skilled in knowledge of how deep their cars would wade began to work their ways across the first ford. The Buick was as old, and consequently as high, as most of their cars. So we wallowed across, taking the fan belt off to keep the ignition dry. The camp was lonesome; only John Mathews and his family and McCoy, finishing breakfast. "When Monk and the other boys seen the rain a-comin' they pulled up stakes for home," John said. "And it rained, I mean."

We walked down to the river together, looking for a likely place to ford. Our stepping stone bridge might as well never have been. We started wading through water just better than knee-deep, swift enough to make a game of fighting to keep from being swept off our feet and carried downstream. We clambered out dripping, went around the next turn, and were confronted by a broad pool.

"Why, you can't even see the quarry," McCoy pointed out. He needn't have bothered; I had already not seen it.

I waded out to investigate the dikes. A few sandbags had toppled from the upper one, but most of both still stood in good condition. Aside

from drilling more holes in the upper layer for blasting, there was just one thing more that could be done for the day. I went back to the Buick for the cameras; I must record the dreary picture.

The river continued to drop, and next day Eaton and Pendley came back to camp. But John Mathews went home for a few days, and three of my people failed to show up, so I was left shorthanded. Down at the quarry the coffer dams were just beginning to show above water. For a while we filled sacks and repaired breaks in the main dam, then went to bailing. We found the dams had created an excellent barrier behind which the river had dropped huge quantities of sand and silt and a generous amount of plain, unclassified mud. Removing this gooey blanket promised to be a lengthy job.

Two days later Monk dashed a pail of clean water against the last of the excavated tracks. It seemed ages since any progressive work had been done on the job. Now, with nearly a full force on the job again, we charged the drill holes in the ledge above Trail Three with dynamite. As everyone raced for the bank, I set off the charge and the thunder of dynamite shook the Paluxy's banks. Smoke rose lazily through the trees as bits of rock and pebbles came sifting down about us. "If that dynamite did as much good as it sounded," Tom Pendley said, "we don't have to move much rock."

The edge of the heavy ledge was completely

demolished, with only the cushion of clay about the tracks remaining intact. The boys all pitched into clearing away the rubble, preparing for the final uncovering. But a stink not related to the fading smell of dynamite hung over the quarry. We "nosed" the smell down to a pile of logs, rubble and trash that had lodged beyond the ledge at the peak of high water. The blast hadn't disturbed the unsightly pile to any extent, but it had shaken it. Monk went to investigate the stench. He drew back and pointed his nose away when he came to the point where four short logs lay jammed together.

"Hey, there's a drowned sheep in here," he choked.

So the boys took time to fire the logs and cremate this victim of the flood.

Sunday ushered in a return to normal times; all morning visitors streamed down into the quarry in unending procession. The crew was off, of course, but I stuck around to answer any questions and to keep an eye on things. There was little chance anyone would walk off with a freshly cleaned track, but on the other hand we'd much rather not have them decorated with graffiti.

It was a better-than-average visitor's day, because I got a new question. Mostly, the pleasant task of talking with visitors on such occasions becomes a great bore because of the predictability of the questions. A garrulous and waggish old fellow threw me a new one: "What kind of noises do you suppose them dinosaurs made? Seems like some o' them forty-fifty-tonners should of been able to leave out quite a beller!"

I gave the man credit for the thought and tried my best, happy to feel safe no scientist of superior standing could challenge me. Few present-day reptiles make noises to speak of, and both lines of dinosaurs are still considered reptiles. But birds, their most probable descendants, are highly vocal. I tried to think of a bird with a long snaky neck whose voice might be comparable. Suddenly I recalled a thought from away back in childhood, the harsh blowing of an avian creature that struck me as being a highly magnified hiss of a snake.

"Well," I told him, "I can only give you a guess, but it will be only a guess. You can't prove me right or wrong."

"I'll take it," he said. "What's your guess?"

Excavating tracks at the Paluxy River site.

AMERICAN MUSEUM SPECIMENS
  A.M. 1   SINGLE GIANT TRACK
  A.M. 2   WALL MOUNT SLAB
  A.M. 3   BASE MOUNT TRACKS
  A.M. 4   SPECIMEN TRACKS

U.T.-UNIVERSITY OF TEXAS
B.C.-BROOKLYN COLLEGE
B.U.-BAYLOR UNIVERSITY
U.S.N.M.-U.S. NATIONAL MUSEUM
S.M.U.- SOUTHERN METHODIST UNIV.

Sketch of the Paluxy River Valley, looking to the south,
showing disposition of tracks inside and outside the coffer
dams.

"Did you ever hear a mother goose protecting her goslings? If you magnified that . . ."

The idea pleased the old man. He agreed if that rasping sound were magnified in proportion to the creature's size, it might be exactly the sound a brontosaur in good voice should make. "But I wouldn't set out to rile one of 'em, just to find out, not even if I was to have the chanst. Ner one of them carnivore fellers, neither. What kind o' sound you s'pose they made?"

"Could be, they bellowed like a bull alligator, magnified, of course."

The old man was pleased with my scientific imagination and urged us to continue the carnivore trail; he'd like to see where the meat eater ate his meal, if he got it.

Monday the crew began to cut the eighteen-inch channel necessary to outline the great slabs to be taken from Trail Three. They worked in pairs at convenient intervals, one man holding a heavy chisel and the other laying to it with a ten-pound sledge. Chips from this operation often flew through the air like bullets, and I thought it wise to furnish goggles for this job. I also had the Glen Rose blacksmith make me up some clamps to hold the chisels

by sort of remote control; a sledge missing its mark and striking the arm of the chisel-holder could cause damage. I had no trouble with the goggles, but it was difficult to get them to use the cumbersome holders for the drills. In everybody's opinion, they were unnecessary and a nuisance.

At the hotel that evening a phone call awaited me from a Y.W.C.A. camp below town. A Miss Thompson, in charge of the camp, wanted to arrange for her group to visit the quarry on Saturday. They would bring lunches, she said, and have a picnic at the fourth crossing. The high point of the day would be a trip to the tracks. I told her how much we would enjoy the visit of her recreational group.

"Then be prepared to have about three hundred of us descend on you," she said, laughing. I laughed back; I thought she was funning, not warning.

I had forgotten the thing by Saturday, and we were all at work on the big slab. Came the sound of many voices. We all looked upstream.

Tom Pendley, a young fellow with a wife and two kids, was the first to grieve. "Why couldn't something like this happen in the days of my youth?" he asked. Headed down the path was an

YWCA group at the Paluxy River site.

180

endless, winding procession of girls. All sizes, all ages, all coming our way. In dresses, shorts and slacks. Plump and slender, blondes and brunettes, redheads and offbreed colors. Three hundred, she said? I don't believe Miss Thompson could count; looked like thousands to me. I laid aside the papers I was trying to work on. Saturday's a poor working day, anyway.

Feminine charm overflowed the bank above the quarry. Beauty threatened to fill the excavation itself. I shooed the girls back from the sight they were hiding from others and out of the way of the workmen. I gave my usual spiel about the tracks, about our task here, but this was no ordinary situation. A clever stratagem occurred to me; I directed them all to the guest book we hoped they'd sign. Naturally this brand new book was not located in the work area.

Never did a clever trick backfire so. At mention of writing down names, everyone thought it a fine idea. A little redhead, dangling her legs over the nearby bank, dropped to where I stood, waving a sheet of notebook paper.

"Mr. Bird, may I have your autograph?"

A hundred other bits of paper came fluttering through the air. Assaying the shower of paper, I was thankful that this was not a student group loaded with notebooks. We'd soon run out of paper.

We soon did. Perhaps even before we ran out, an inventive miss picked up a chip from our flaking operation, for my autograph. What could be more appropriate than the field superintendent's own name on a bit of rock that came from the site? Since this was not very good for my fountain pen, I soon abbreviated my signature to R.T.B. When disappointment was expressed over this, I added "to" and the girl's name. I recall the first one was Sylvia Waters. I wonder who she is now. When the affair was apparently headed for a race between writer's cramp and the total destruction of my fountain pen, Miss Thompson saved the day. "Let's all go down to the crossing for a picnic," she interrupted. "You must be our guest." The picnic area was sodded; no more stones. Anyway, stones from a mere picnic area would have had no relevance.

At the end of a pleasant if not highly productive Saturday afternoon, Miss Thompson began the task of getting her girls together.

"I want to thank you for your time and for your talk. I hope we haven't been too much trouble," she said.

Channeling of trackway slabs nearly complete at the Paluxy River site.

181

"Trouble?" I asked. "What trouble? Compared to the flooding we've had, you've all been sheer pleasure. I've had a wonderful time. And it's sort of fun to be the hero of the hour, as long as it's for only an hour."

Miss Thompson laughed. "That's what you think," she said.

"What do you mean?" I asked.

"This bunch goes back tomorrow. I'd like to bring a fresh group just like this two weeks from now. I'll ask them not to bring paper."

Following such a glorious day, it was sort of nice to get back to the grubby job at hand on Monday. The work of chiseling the long channels through the track-bearing stratum was hard and dull, but it was nearing completion. The channels looked like ditches for water pipes cut in solid rock, and it was a relief to know the end of the job was in sight. We had now nearly freed from the matrix two great rectangular islands. The one for the American Museum was twenty-nine feet long by eight wide. I could already envision it set in the base of the *Brontosaurus* mount, with that great sixty-seven foot skeleton stepping out of these tracks. The many thousands of pounds of fossil bone, the framework of geology's most gigantic quadruped, the very symbol of dinosauria, would seemingly have left his tracks in the soft mud, a frozen symbol of life and movement such as no dinosaur had ever been before. It would rival the superlative elephant mount in the entrance to African Hall. And Brown would say . . . The moment was hard to wait for.

I had made a diagram of each slab, so that the positions of the sections might be recorded as removed. In the afternoon I started a similar task for the block going to the University of Texas. The clanging sledge hammers resounded in the walled bed of the Paluxy, ever so faintly accompanied by the *sotto voce* undertone of the rustling live oak trees on the bank. Steel against steel, steel against rock, a hypnotic monotony of sound. The men were paired up, two by two, changing now and then between wielding the sledge or holding the chisel. No conversation, except as they changed positions. There was a curious break in the flow of sound . . . and the sound, all sound, stopped. Looking for the cause of the off-key note and the following silence, I saw the seated Moxon fall to his face beside his chisel. Monroe Eaton stood over him, half-leaning on his sledge, his face white with horror.

What happened was too clear. There was no clamping holder on Moxon's chisel. The sharpened car axle lay where it had fallen from his limp hands. Somehow Eaton's sledge had struck a glancing blow that spun the heavy head against Moxon's temple. The force of such a blow, delivered from the end of a three-foot handle, would have been tremendous.

The group stood frozen. I was sure the stout, sandy-haired little guy was dead. A white spot on his temple became an ugly red, and blood welled from a break in the skin to drip slowly onto the rock. Eaton straightened slightly, rubbed a hand across his forehead, pushed a tumbled lock of sweat-damp hair out of his eyes.

The tension broke. Men bolted toward the fallen Moxon, except Manuel Gosset, who grabbed a pail of drinking water first, and poured it on his head. We stood about, silent and watching, until Moxon opened his eyes.

The united sigh of relief was the best breeze in the river bottom all that day. Then someone laughed nervously.

"God A-mighty, he's got a tough skull! I mean a real tough skull!"

Eaton said through dry lips, "It's a wonder I didn't kill him. The handle twisted when I hit the chisel, and it got away from me."

Moxon sat up, holding his soused head in both hands.

"How do you feel, Mox?"

"I . . . I guess I'm all right."

We helped him to the bank and to a shady spot under the live oaks where we bandaged his head as best we could. A great lump rose under the red-stained bandage, and he complained of a terrific headache, but otherwise he seemed none the worse for the experience. The boys waited on him tenderly until it was clear there was no more to do. Then there began a silent movement toward the bushes . . . to retrieve discarded chisel holders. Moxon went home feeling pretty tough, but Gosset, who accompanied him, said no complications were resulting from the glancing blow.

Next afternoon we were ready to start taking up the brontosaur slab. I showed Eaton how to apply a plaster and burlap cover to the edges and upper surface of the first foot or so of the slab. Mr. McCoy stood by, ready to give assistance. John Mathews and Tom Pendley waited with mattocks.

The channel at the rear end of the slab had been widened to give the men a place to stand. They loosened the clay under their feet and shoveled it out. Then they began to cut back under the

Carrying pieces of trackway slab uphill at the Paluxy River site.

Little Tommy Pendley taking an impromptu bath in the hindfoot track of a brontosaur, Paluxy River site. Photo by Bird. Courtesy Department of Library Services, American Museum of Natural History.

block. Presently the end sagged slightly, breaking along a convenient fracture. Mathews and McCoy stood ready to ease its fall. Monk slid a sandbag under it at the last moment. The freed mass weighed so much that it took all hands to stand it on edge, but it could be lightened considerably by chiseling six to eight inches from the back. I turned to McCoy, a handy man with hammer and chisel.

"Mr. McCoy, I'm going to delegate this work to you."

Monk plastered the edges of the new, clean fracture on the main slab, as the men with the mattocks began to undermine another section. McCoy finished reducing the weight of the turned section, and Monk plastered the edge that had made contact with the block.

I painted a number on the dried plaster jacket and sketched in the outline of the section on the chart. It had been late in the afternoon when we started, but we got another block turned and plastered before quitting time.

"What bothers me now," I told the men, "is getting these to a place where we can pick them up by truck. I guess we'll have to build some sort of a cradle and tote them up to Mr. Wilson's field."

The field bordering the top of the thirty-foot bluff above us was the closest to the quarry of any spot we could reach by truck. Next day I fastened two eight-foot planks together with two-by-fours and heavy spikes. Even with their jackets on, the blocks couldn't be treated as rough stone. We placed this stout platform in the cavity beside the prepared sections and gently turned them onto it. Four crowbars were slid under the platform, and eight men at a signal heaved on the bars, inching the bars along a little path up the thirty-foot embankment. The ancient Egyptians who built the pyramids could but have marvelled at how little things had changed.

Days passed like this. Moxon returned, head bandaged but able to tote his share of the daily load. By brawn uncluttered with modern engineering (after all, we were working in "dinosaur time") we made good headway. We plastered a number of blocks and wrestled them, a few at a time, to the top of the bank. The threat of rain was ever with us, and I didn't leave plastered blocks in the quarry overnight if we could help it. The water would surely soak away the bandages and, worse yet, obliterate all identification numbers. The blocks at the top of the hill, awaiting a trip to town, were always covered with canvas.

Our publicity had snowballed to the extent that *Pathe News* thought the quarry of national interest, and one of their cameramen spent a day making newsreel shots.

It occurred to Pathe's director it would be a good thing to have a child in a picture, as a human interest thing. The families had of course brought their children along to the job site, and all of us who had had them underfoot all this time could attest to their human qualities. I was appointed to select one on the basis of photogenic qualities. Mrs. Pendley suggested her Tommy might not fill the bill better than anyone else, but since he was in sight he might be easiest to catch. She called Tommy over, and I suggested he pose taking a bath in a dinosaur footprint. Little Tommy was human enough, but as to taking an unscheduled bath in the daytime and in front of everybody at that, he wasn't interested until I told him he could keep all his clothes on and splash all he wanted. And he might be the first human in all history to have a dinosaur footprint for a bathtub. A three-year-old is seldom motivated by consciousness of his place in history, but being allowed to splash around in the water in all his clothes won him over. Both the Pathe man and I shot the dickens out of Tommy Pendley, Junior, and the years between lead me to believe it was the greatest human interest shot ever made in scientific surroundings. Tommy's mother also played a bit part; she was the visitor to whom the resident scientist explained the occurrence of the track. Neither Mrs. Pendley nor I ever won picture contracts for our brief bits, but Tommy Pendley in his first-of-its-kind tub was widely used.

Threatening clouds did a good bit to speed up our work next day; they gave very muddy wings to our feet in getting the last of the prepared blocks out of the river bottom and under canvas up on Wilson's field. Thunder grumbled at us from just over the nearest hills, reminding me that some of the channeled tracks, ready to take up, might be dislodged and the sections scattered, if the force of the impending flood should be strong enough. I set the boys to putting sandbags about them, and Pendley and Eaton went to work reinforcing the main dike wherever it was weak or low.

A wind stirred in the leafy live oaks. All afternoon it had been hot and sultry, but the new breeze was cool and smelled like rain. The sky darkened, and a few drops spattered loudly on the hot rocks. A vivid flash lit up the sky in the direction of the fourth crossing, its bright band of fire reflected in the river. I thought of Brown and of the nights we had struggled with the tarpaulins at Howe Ranch Quarry.

"Dogged if that there lightnin' weren't close by!" Mathews barked as he looked up at me.

Rain began to come down steadily. The drops speckled the backs of the men's shirts. "Let's move it, boys," I called out. "Gather up the tools, and we'll hide out in the Buick 'til it's over; it's only a short shower."

The men finished the job of padding the blocks, as McCoy scurried about gathering tools. In the distance, from the path approaching the quarry, came a child's piercing shout.

"That's one o' your kids, Mathews," someone called out. "Sounds like something's wrong in camp."

Everybody listened, but the only sounds were the wind shaking the leaves and the patter of raindrops. I climbed the bank above the quarry and caught sight of one of Mathews's little girls dashing like mad in our direction. Monk climbed up on the bank beside me. We heard the child call out again but lost her words in the gusty wind. She disappeared, then topped the next rise, her legs going like pistons. She shouted again, a high-pitched, childish scream, "Mr. Bird's Buick! Mr. Bird's Buick's on fire! The lightnin' hit Mr. Bird's Buick an' it's afire!"

Loading pieces of trackway slab onto the truck at the Paluxy River site.

35

I STOOD LOOKING at the Buick in the falling rain, thankful that she had not sheltered us from this particular downpour. Lightning was only a passing event in a life of many trials. Eleven years old, she had been driven so many miles and fleetingly dammed so many rivers it seemed foolish to throw away stuff as rare as money on her. If it had been practical, even at the start of this expedition, to replace the canvas of her touring top, to touch up or to repaint the light green body with its dark green trim and replace some of her more intimate parts, I should have done so. Now her very dilapidation set her apart and gave her color, even the rust spots. She symbolized all that went with fossil-gathering the hard way. If she continued to run and managed to last out the season, I should be satisfied and felt Barnum himself could ask no more of her. A gaping hole in the right rear corner of a long-ago jaunty touring top, where the canvas was blackened and charred by the bolt, marked the spot where it had found metal.

"She sure would of burned up, if we hadn't beat the fire out, what with the rain to help us," Mrs. Pendley told us. We believed.

I got inside, out of the downpour. There was little shelter left under the remains of the top. A smaller hole than the new one poured water onto the leather back of the front seat. This lesser rent was a memento of a sizable hailstone in Michigan, coming home from last year's jaunt. I moved behind the steering wheel, the driest place I could find, and watched the rising storm lash the surface of the Paluxy like whips; in just a little while the restless stream would be booming with a fresh rise. The water would come in over the quarry dikes again, mud in tons would settle in the great eddy above Trail Three again, it would be days before we could go back to the work we had left in such a pell-mell hurry.

In a lull in the storm I called out to a lanky new boy standing in the shelter of a nearby tent. "Elmer, if you or any of the boys want to get back to Glen Rose, now is the time before the water rises at the fords."

He piled in, tried to hold the edges of the hailstone wound in the top together with his fingers, and motioned to two other boys who piled in the rear. Momentarily they wiped dry spots to sit on but could do nothing about the lightning's recent addition to the top.

The car wouldn't start. A hasty and rain-drenched diagnosis indicated a clot in the gas line, possibly brought on by the bolt of lightning. I took the top off the carburetor, siphoned off a bottle of gasoline, gave the bottle and the siphon hose to Elmer.

"Elmer, if you ride the fender and dribble the gas to her, we'll get you to Glen Rose just like I said. A little wetter than sitting under that hailstone hole, but . . . And we'll put the hood in back; it'll keep the rain off the laps of the rear seat passengers."

And away we went. Elmer, wedged in behind the spare tire, did a great job. So did the old Buick. We rolled into Glen Rose as proud and happy as we were wet.

Four days later we made our way back. Once again, we shoveled the last of the mud out of our quarry. Fortunately, disposal of the mud was simple: the same river that left it with us took it away. The main dike hadn't been hurt much, though a good bit of the upper dike, catching the full brunt of the Paluxy, had been partly toppled.

We were able to get the situation well in hand before the next storm broke. The season's worst, it left all our diking in a bad way. Worse than that, it broke all our backs, spiritually. Wilson hadn't half told us about his river. I was growing worried.

Certain things were expected of me, with little allowance for the whims of a river. There was still a great deal to be done, and I couldn't afford to throw manpower into what seemed to have become an endless game of mud-flinging.

So when next we went back to clean up, I prepared for the defense. I took along movie cameras to record the story of some of our clean-up problems. As I told John Mathews, "If repeated floods keep us delayed much longer, I'll want some solid proof what we've been up against."

I set up the movie camera to shoot Monk in knee-deep mud, bailing away like crazy in the rich brown gravy. I ran a few feet of McCoy, up to his fat knees in what had been one of the channels around a slab, digging away to relocate its position. I shot Sam Mathews and Tom Pendley working on a new inner dike, dragging muddy sandbags into position, swinging others into position on the new wall. The new dam, enclosing only the part of the quarry in which we had present interest, promised better protection for our increasingly precious slabs and their tracks.

Suddenly, a strange rumble filled the air. Not the wind in the trees, not the normal rush and rumble of the river. I looked upstream, toward the source of the odd sound. A roll of water in a wave nearly two feet high tumbled toward us. It stretched from bank to bank, advancing on the face of a black rise. I watched the weird sight in horrified fascination, watched tree limbs and logs tumbled like match sticks. The men heard the sound too, looked up, stared stupified. Incongruously, an old mattress rose through the roiling mass of water and rode the crest of the churning wave.

The water outside the main dike began to rise, and we came to our senses.

"Boys! Quick! Sandbags! Get sandbags around those tracks!"

Monk and Pendley leaped into action. Others jumped to aid them. I jumped too, and nearly fell over the movie camera. The moment was priceless. I fixed the view finder on the rolling wave sweeping past the place where the upper dike had been. Film raced with the racing water. The river level outside the new dike raised a foot in the twinkling of an eye. Water boiled over the low places and began creeping up the wall of our new inner dam. I swung the camera to catch the frantic movements of the men. The brontosaur slab, so freshly cleaned of mud, swarmed with frantically moving forms. Water topped the inner dike, poured down into the excavation, swirled into the big tracks in a dozen

places. Meanwhile the sun was shining brightly, doing its very best to make the scene ridiculous and incredible.

What had caused the flash rise? We remembered the dark clouds of the night before in the west. It hadn't rained in Glen Rose, but a cloudburst must have hit the river's upper watershed. We learned later this was an occasional dirty trick of this rambunctious little river.

At the third crossing the river was wholly at peace, just a quiet little treacherous river in its summertime doldrums. We waited here a bit in the warm sun. When we saw the tumultuous wave coming our way, we drove the Buick across first and then laughed at it. We repeated this at the second crossing; there was nothing more to do for the day anyway. The pictures at the quarry seemed somehow to make up for the trouble this rise in the water would give us.

The Fourth of July came, hot, still, dry. The long heralded and slightly retarded summer drought seemed at hand. The river was down, and most of the men agreed to work on the holiday; they had missed so many days in June that they were glad to make the extra time.

The days wore on. More than seventy marked sections of the slab, containing the first footprints of a sauropod dinosaur ever to be cut from rock, had been removed from the quarry. Each day we plastered the blocks until we had enough of them prepared to carry to the bank.

Beside the steep path a tree stood slightly in our way. One day McCoy brought an axe out from camp, and while the rest of us went to work in the quarry the woods rang with the sound of McCoy's blows. When Monk went up near the trees for plaster, we heard him shout, "Hey, Tubby, if you're gonna cut that tree down, why don't you cut her all the way down? What's the sense to cuttin' her off waist high?"

The chopping stopped. McCoy laughed. "I ain't aimin' to cut her to the ground. I want to fix us a stump to rest the plaster pieces on, when we tote 'em up."

An hour later we began moving the day's plastered lot of blocks. The platform was brought down into the excavation and the first load rolled onto it. The men chose lifting partners. I took the end of a bar across from Pendley, who was about my build. We lifted our load and started up the cut we had made in the bank to reach the high spot among the trees. The load was heavy, the path winding up the bank steep. When we reached

Tubby's stump we rested the middle of the platform on it, and relaxed with our burden. The pause refreshed everybody, and McCoy was jubilant.

"See there? My scheme's a good 'un after all! It just ain't half the work to come up the hill when we got us a breather!"

"This only goes to show," they told him agreeably, "you stick to a job long enough, you're bound to do some good. From how on, this spot will be known as Tubby's Rest."

A letter from Brown announced he wanted to spend a little time in the Big Bend area of Texas, looking for dinosaurs where Don Guadagni had found his palm leaf, in the upper Cretaceous. Erich Schlaikjer was to go along with him. For working in the field Barnum planned to use the Ford truck still stored in Rock Springs. Erich was to go out to Wyoming, drive the truck south, and meet Brown in Glen Rose. After a look at our quarry, they would go on to Marathon and Big Bend. Then, as soon as I was finished with the Paluxy, I would join them.

Erich arrived about mid-July. That same afternoon, I met Brown at the Dallas airport. He had heard about the old Buick's latest bout with Fate, but only when he saw the lightning-ruptured top did he realize how it emphasized the old car's advancing infirmities. The sight of her standing by the curb as he left the plane must have been to him like an old plug horse standing on three legs. His eyes reflected amusement, surprise; then he laughed as I had never heard him laugh before. I explained there was no place in Glen Rose to have the top repaired. We climbed in, and I pressed the starter.

"She's getting so she doth protest too much," I pointed out, as the old girl spit, sputtered and coughed, and then with a puff and a couple of sneezes settled down to business, letting us know she was ready to go to Glen Rose. Brown laughed until he shook the car more than did the complaining motor. An hour later we chugged and wheezed into Glen Rose, picked up Erich, and rolled out to the quarry. As always, Brown's enthusiasm was an inspiration. He moved up and down the line of tracks, laying plans for me to collect more specimens for the wall collection. He spoke of Dr. Charles Gilmore, of the U.S. National Museum in Washington.

"I've written Gilmore, and he thinks he could use a block of tracks," he said. He also thought the American Museum and the University of Texas ought to have a track apiece from the exceptionally big trail Eaton had discovered.

Erich came up with a proposal. "And I'd like one of the smaller ones for Brooklyn College."

The summer wore on in blessed dryness. It grew hot, and both banks of the Paluxy bounced the heat down on us and kept any breezes out. Balancing saline tablets against sweat, the men kept to the job. Grass along the river grew dry as dust, and the often angry stream that tossed trash twenty feet above our heads shrank until we could have handled its flow with a bucket brigade. Leaves curled up on the trees and began to fall, warning us summer was coming to an end. Only cedars and live oaks remained green, and the willows, of course, with their roots drinking directly from the river. Sometimes dark clouds peeked over the horizon at us, but they went away.

So we finished with the brontosaur slab; it lay now, a great pile of white blocks on the bluff above, in a corner of Wilson's corn patch. When he cut his crop, I got permission to break a road across his acres to haul the blocks to town. I had removed the trunk from the back of the Buick, and we could load a three or four hundred pound block in there, with smaller blocks piled on the floor. Night after night she groaned and puffed her way to town, unburdening herself at the local lumber yard. At the same time trucks began arriving from Austin to pick up the sections of the slab for the university.

Toward the end of August, a long distance telephone call came in from Brown. Erich's vacation nearly over, he must soon go back to Brooklyn College. Brown, never having learned to drive a car, would be unable to get back in the field unless I joined him. The end of the long job in the river quarry was in sight, but so was the end of the dry season; what could I promise?

"With good luck," I told him, "I might be able to get to you about a week from Tuesday."

"Have all the luck you can; I'll be looking for you," he said. "Erich and I have seen little in the way of dinosaurs, but I have found part of a giant crocodile skull."

The long distance line rasped and sputtered, and Brown went on. "We won't be able to tell much about it until it's worked up in the lab, but the teeth are the largest of their kind I've ever seen."

At quitting time Friday the sky was cloudy. A dark cloud bank came on in the afternoon. I walked out of the quarry with the men, leaving at least a dozen plastered and numbered sections lying in the hole. I hadn't grown callous about threat of rain, but handling sections in mass lots was by far the most effective way of using manpower. I had taken such chances often of late; this was just one more chance.

In the night, it rained. The first fall of many weeks was heavy, and the rise that followed was in keeping with this.

For the next two full days, we had no problem at all; we couldn't get to the job. On the third day we fished some of our blocks out of mud, re-plastered and renumbered them. The next day we just cleaned mud off of the remaining blocks and did the same. It was too late for us to learn a lesson from this; this was the end of the run.

When Eaton laid aside the last plaster bucket from the last plastering job, I got out the cameras to bring the story to a close. It seemed a geologic era had passed since the day in May when this had all begun. We had removed over forty tons of rock from the river bed. We had used a ton of plaster of Paris and eight thousand Kleenexes instead of rice paper. At one time nearly five hundred feet of quarry had been exposed to view, to my knowledge the largest dinosaur track quarry ever opened. Fourteen men had worked for me from time to time on the crews. The eight boys who carried the last load from the quarry passed in front of the camera and out of sight.

The great pile of plastered rock destined for the museum lay in the Glen Rose lumber yard to be crated. That, I decided, could wait. I had fifteen hours to get some sleep and drive to catch up with Barnum in far-off Marathon.

36

THE AGED BUICK droned across the endless expanse of west Texas. The pace of the last few days had left me with numbed nerves, and I guided the car toward the ever-receding horizon in sort of a trance. Late Tuesday afternoon the wheels rolled me into Marathon. At the Gage Hotel I caught sight of Barnum coming down the front steps, dark as an Indian from his weeks in the desert sun, in smiling good spirits.

"You missed Erich by just a couple of hours," he said.

I drove the Buick into the hotel garage to put her in storage. The weeks of using her as a stone-wagon in Glen Rose had brought her ever nearer to ultimate collapse, and I backed her in tenderly; if I bumped the wall, she might well settle down dust unto dust like the "Wonderful One-Hoss Shay." Walking away from her, I looked back only once; I had a premonition we'd never meet again.

The Ford truck outside was covered with dust. It, too, had seen heavy duty of late, but the body and frame under the brown film looked sound and reliable. Loaded in back were all the accoutrements of overnight camping: a mess box, a couple of cots, bed rolls, a large canvas tarp. Also the usual fossil-digging tools, bags of plaster, and a bale of burlap sacks. A drum of gasoline. Two empty drums, and three milk cans. Brown appeared around the corner pulling a garden hose.

"We'll need our own water in the Big Bend," he said. "Stick this nozzle into one of those empty drums, and we'll see what happens."

When we had filled the drums and the milk cans we drove around to Marathon's grocery store to stock up. By nightfall we were ready for the road again; Barnum wanted to make an early start.

By eight o'clock we had bumped across the Southern Pacific tracks and were headed south. It was good to be back behind the wheel and headed somewhere on purpose. The gravel road pointed toward an empty desert and distant peaks. It climbed a bit and snaked its way through low hills with limestone ledges covered sparsely with non-descript brush. The light grey rocks were Paleozoic in age; no dinosaurs. We eased up the slopes of the Santiagos, and from their top we could see the distant Chisos off in the Big Bend country. The horizon had moved back in all directions.

"The Lord had batter to spare when he laid out this country," Brown remarked.

The formations in which he and Erich had worked were Upper Cretaceous. When we came upon the first outcrop, rising with the uplift of the Chisos, Brown pointed out a low series of sequestra not unlike some of the upper Mesaverde close to Rock Springs.

"Those are the Aguja beds, with the Tornilla Formation above them, in the distance," he added.

Bands of blue and purple clays, with inter-bedded grey sandstones, lay above the rolling brown sequestra. So far most of their finds, Brown said, had been in the Aguja.

"The bones we've found have been scattered, often broken, but the exposures are good and easily checked out."

I listened again to the story of finding the big crocodile and was sorry I had missed out on it. Noon came, and we had lunch at the end of a small side road, where Brown told me he and Erich had set up a tent, and where Erich had caught a rattle-snake. We lingered at this base camp only long enough to pick up more equipment, then swung east, into the country drained by the Tornilla Wash. The terrain dropped toward the Rio Grande Valley, and the road grew rougher. Except for a narrow band of brush along the washes, vegetation was thin and sparse and only emphasized the lack of cover from the glare of the afternoon sun. Near evening the trail led into low hills and crossed a long, flat bed of sand that was the Tornilla Wash. The sun set behind the Chisos, and night came on quickly. The bumpy road became a deserted trail that turned and plunged into a tangle of brush in the dark. Brown directed me down a little draw with no sign of wheel tracks ahead. The headlights bored into the rising flank of a soft clay hill.

"Let's look this thing over afoot," he suggested.

I stopped the truck. We seemed to be on moist ground, for we were surrounded by clumps of brush so thick that in the unlit gloom they seemed to be squatty trees. Gnarled catclaw and dense stands of cactus that sometimes towered over us made up the thicker growth, while spindly ocotilla stood about on open ground, the long and nearly leafless stems bobbing about like wriggly strands of barbed wire. We got out our flashlights and found our way around the clay hill. An open space on the other side seemed to indicate the direction followed by the one-time road. We continued along it, exploring. Barnum shook his head.

"This isn't quite the spot," he said. "Let's get back to the truck; it's just a bit ahead."

We drove on just a bit, to the first level place, parked, built a fire of dead catclaw and cedar. The flame flickered and winked, crackled and sparkled and spat sparks. It played jerkily on the rises of nearby sequestra and against the surrounding desert growth. Supper was served on the table-like lid of the mess kit let down over the tailgate of the truck. Stuffed with good camp cooking, we swung our canvas chairs to face the fire, and Barnum lit his old briar pipe. The evening passed quickly, with Barnum in his finest story-telling form. I had heard some of the tales on other occasions, but the new telling carried the same old savor. When our pile of gathered fuel ran low, we crawled into blankets almost too warm for the night.

In the morning we drove to a point nearer exposures of the Aguja. Brown pointed out the horizons most likely to carry dinosaurs, and we went to work on a long, open ridge not far from Tornilla Wash. The flat bed stretched before us as though the sand, water-swept in undulant and winding bars, was itself a flowing stream.

We both came upon bone fragments, scattered over the face of a sandy exposure, weathered grey from lying long in sun and rain. Some pieces were fairly large, identifiable as bits and pieces from ribs and sections of limb bones. I picked up a piece from a caudal vertebrae. After enough of this, Brown said disconsolately, "I've almost forgotten what to do in case I should find a whole bone."

What had caused the mass breakage and the odd lime deposition over the worn and rounded ends of broken pieces of bone? There was no plain answer in sight, no clue in the faces of the surrounding sequestra; the sands and clays that faced us might have been Mesaverde. The only likely explanation lay in the violence of a strong river or a heavy surf along an ancient shore, that had broken and swept these bits to this final resting place. Brown located one sizable concretionary mass, heavily lime-encrusted, which encased, in a questionable state of preservation, what appeared to be the flat skull of one of the armored dinosaurs.

By lunch time, the dry air was blistering. Mountains boxed in the area, so breezes had no chance to blow. Had they blown, they would have been blast furnace breezes.

Back at the truck, we refilled ourselves and our empty water bag from a ten-gallon can, and Barnum dressed his shining dome with a wet handkerchief. When I complained a bit about the heat and the shining promise of more heat, he grinned.

"You should have been here in August."

We stretched our canvas tarpaulin over the top of the truck and anchored it out on one side for shade. We ate, wetting down our lunch with generous amounts of hot tea, and stretched out on our cots under the canvas. The steaming tea got us to sweating profusely, and we were soon resting as comfortably as conditions permitted. We guessed it was about a hundred and ten in the shade and were happy we had no thermometer to prove it out. By three o'clock the worst of the heat was over and we went back to sweating in the sun. The approach of darkness signalled time to find a new place to camp.

By mid-September we had run through the last of the Aguja that was at all accessible by road or trail. Brown had added little to his meager collection. He got out his maps to speculate on unexplored bits of Aguja marked on the map as the "Banta Shut-In." With a name like this, we assumed it wasn't on a main road nor much of a metropolitan center. It might even be hard to get to. Standing at the base of the Chisos and looking across the rough desert to the distant beds, we agreed the only way to the Banta Shut-In, if any, was via the Tornilla Wash, which ran through the area. The map showed a stretch of this wash between Boquillas and the place we wanted to reach. Boquillas is a dusty little cluster of adobe huts on the Mexican border, where the wash dumps its litter into the Rio Grande. About eight miles up as the crow flies—a bit more by way of twists and turns and meanders of an ugly little mountain drainage ditch—was the place we thought we wanted to look over.

An ordinary machine might have been unable to do the job, but the Ford truck had understudied the old Buick for so long and over such rough trails that our trip up the wash was not difficult. Still, it was sort of special. The wash was a narrow, high-walled canyon, intrusive rock with vertical fractures that made for high walls and sharp turns.

"As a recent expert on floods," I told Barnum, "I know this is where we don't want to be at the first sign of rain anywhere around."

"I know what you mean," he said. "Any sign of a wet-looking cloud, and we head for Boquillas a lot faster than we're leaving it."

Compared with prospecting we had been doing, Banta Shut-In was pleasant. The high walls with nearly sheer drops to the bed of the wash meant that for a good bit of the time we could stay in the shade, guiding our coverage by the time of day. When we did find the sacrum of a ceratopsian exposed in a ledge at the crest of a series of sequestra, Brown broke out a beach umbrella he had in the truck, and we set it up, propped by rocks, so we could both work in the shade. Happily, the piece was set in a soft sandstone matrix, ideal for quarrying. The site was inaccessible by truck, but Barnum said he had a plan for getting it out if we got it loose.

We managed to make this pleasant work-spot last for days, as fine a pack of days as we had ever spent quarrying. At week's end, we had the lump loose, and it was time to go for water from the Green Ranch outside of Boquillas.

We got to Boquillas in late afternoon, and it was coming on dark when we got to Green Ranch. Brown had met the Greens earlier in the season and had established a friendship with them. Without telling me much except that he was acquainted with them, he went to the door and knocked. No reply. Instead of leaving the doorstep, however, he fished down beside the sill and came up with a door key. He opened the door and motioned me in, a twinkle in his eye.

"Mrs. Green told me to make myself at home anytime," he said.

We went into the kitchen, and Brown immediately busied himself about getting a good supper on the range. After supper we washed up the dishes and were just preparing for bed when the Greens came home.

"Will you have a bite to eat?" Barnum asked. "It won't be a bit of trouble. And welcome home; we'd hate to have turned in before you got here."

When I saw how pleased the Greens were to see us, my nerves settled down a goodly bit and we had a pleasant evening.

"We want something to make a little cart, a hand cart, out of," Brown told them, "to get a lumpy specimen down off a ridge. A couple of old boards is all, and maybe a couple of wheels or rollers."

Mrs. Green smiled. "Must you have them before bedtime, Dr. Brown? Or will tomorrow do?"

Brown opted for morning, when we scrounged the door of a junked car, a length of inch-and-a-half pipe, and two pulley wheels from a piece of discarded farm machinery. We soon strung this all together with a few odd bolts. Back at the Shut-In it took us two days to finish the job, but the actual process of trundling the plastered pelvis down to the waiting truck was accomplished fairly quickly. The only difficulty was in keeping our cart from getting away from us on the way down.

We rolled into the Banta camp with the big white block in the back of the truck just before sunset. Across the river, in Mexico, the great mountain, El Carmine, was only eight miles away. From where we were camped, the Rio Grande, in the bottom of its canyon, couldn't be seen at all, but in the clear air it seemed we could reach out and touch the sheer, bare, dramatic red mountain. As the sun sank behind the Chisos at our backs, reflected light flooded the wall of pale grey limestone and turned El Carmine first rose-pink, then flaming red as though lit from within. Minutes passed, shadows deepened and Tornilla Wash grew grey, but still the mountain flamed, reflecting every hue of the western sky.

By mid-October, it seemed September had never been; the passing time took with it the terrific heat, and prospecting became a pleasant thing again. No passing clouds made us wonder if we must cut and run for Boquillas to save the truck. It was a nice time.

It was nice, too, to recognize that we were

more or less wrapping it up, that the time was drawing near to head for home. Barnum and I both had a great preference for field work over lab work, but for several reasons hunting fossils isn't listed anywhere as a winter sport.

We had by now fairly exhausted the Aguja. We had done a good bit of scouting through the local Tornilla Wash and dabbled a goodly bit into the marine Boquillas. Both formations had enriched the season's collection of vertebrates—nothing spectacular, but the bits and pieces added up to a respectable amount.

Some of the marine material we uncovered, while it lacked the excitement of dinosaurs and early mammals and such, was on the spectacular side. Crossing a dirty brown hill one day, we noticed the ground under our feet looked like clay, felt crisp and gritty, crunched under our heels like dry bones. I picked up the largest piece of the material I could find on the surface and noted one side of it showed mother-of-pearl. We dug beneath the sand-like surface and found we were walking over a fantastic oyster bed. It covered the crest of the hill we walked on and shaded off into a flat beach. Digging through the loose rubble, I turned up broken pieces of what had been an oyster shell fifteen inches long. Brown said, "Let's trench this a bit to see if we can't pry out a few entire shells. These aren't an unknown type, but they are unusual enough to be worth collecting a few."

This had been an oyster bed in Cretaceous times. Shells were ranked solidly side by side, lips up, and the cavity that once contained the living part of the oyster held almost all of four fingers and the palm of my hand.

A week or so later, on the west side of the Chisos, we left the truck to look over a long, low bench of Boquillas limestone which paralleled the road. We separated to walk on converging paths back to the truck along the bench. The first few yards proved barren, only a flat layer of limestone. Then I came upon a flat oval shell, so huge I couldn't at first believe my eyes. A fossil clam shell as long as my pick handle. I looked around for one completely weathered out, but this was too much to expect. The outcrop of shells and limestone continued toward the point where Brown and I were to come together. He grinned at my delight over my discovery. "This isn't a major discovery," he said. "But they're uncommon enough to be worth taking a few along."

When Brown came to the conclusion we had

exhausted the surface potential of the Big Bend region, he turned to his maps again for inspiration. From Presidio, a dirt road followed the Texas bank of the Rio Grande upstream. Eventually this wandering through a remote region would lead us to the tiny town of Valentine on U.S. 90, not too far from Marathon.

"We'll make this trip sort of a vacation," he said. "I think we have one coming. We'll just look over the rocks along the way in this new country while we're resting."

We spent the first night in the little town of Terlingua, centered about a quicksilver mine. The mine itself wasn't much of a producer but was able to hang on by showing visitors through the operation, because mercury-mining is fairly rare in this country. We gathered up a few mineral specimens from the region and decided what to do next on our vacation. "The Grand Canyon of Santa Helena" was a name on the map that intrigued us. Just a bit off the route to Presidio. Still, Barnum felt we might as well drive down to see what there was to see. The dirt road was rough, winding across flat desert dotted with irrigated fields.

The fact that the Rio Grande didn't look like the Rio Grande didn't bother us; it seldom does. The Canyon of the Santa Helena, however, was a different story. We came upon it somewhat suddenly, through the cottonwoods, coming at us as a great split in a rising cliff.

The river flowed from the mouth of this great red canyon. Constricted here, it was faster and deeper than is its usual puny and lackadaisical style. Downstream a bit, once out of the canyon, it became itself again and broke out with a series of shallow rapids. A fisherman went by us in a rowboat propelled by an outboard; he had just been setting out lines. He pulled up at a little dock on the bank under the great rising red wall. Another man was there, with another small boat. The road ended just a bit short of this rock, backdropped by the sheer straight walls of this red canyon.

"They don't know it yet," Brown remarked, "but one of those fishermen is going to get talked into running us up that canyon a bit."

The man in the first boat left before we could make a choice. The other fellow gravely explained he would be seriously stretching a point by carrying passengers on waters of an international boundary. When he had convinced us of the gravity of the situation, he added, "But I ain't seen none those sonny-beeching border patrollers aroun' here for days; you wan' to have the chance . . ."

So we piled aboard with our picks and collecting bags and violated international law right and left. For some reason it always seems harder to believe you are looking at two different countries at these natural boundaries than when you simply step across an imaginary line on the ground. The tall red cliff on our right was U.S.A.; the opposite cliff, identical in all ways, was Mexico.

"Let's push over to the left wall," Barnum suggested to the boatman. "Run in close and go slow. Maybe we can knock off a few Mexican chips and fossils."

The flanks of the canyon were Fredericksburg limestone, fairly well loaded with shells. Sometimes we stood up in the boat and chipped away at the cliff on either side; sometimes we stopped at little sandbars along the way. In an hour or so we came around a bend, grounded on a bar in more shallow water, and the boatman decided we were becoming too exposed, so we headed back. I labelled our samples: Cretaceous, Fredericksburg ls., Mex. & USA.

On our third vacation day, we left the rough river road for a road over which cars hadn't been driven for many years. It led east, into a region where Brown had read in a scientific publication that there were dinosaur tracks to be seen in the neighborhood of the Hat Ranch. Little detail was given in the article, but since it was in Cretaceous country it was possible, and we felt we should check it out. More than most dinosaur people, Barnum and I were intrigued by dinosaur tracks.

We prospected a few spots on the way, but for the most part only assured ourselves the area was wrongly labelled Cretaceous. Barnum came back to the car from one short jaunt with a limestone-encrusted mammalian bone to make his case. It was a tibia, obviously mammalian, from a little animal about the size of a dog. It could only have come from a creature a good bit this side of Cretaceous time. "I wonder what this will make of those dinosaur tracks we are going to look for," Brown speculated.

We kept on toward Hat Ranch, nevertheless, with renewed curiosity. The trail, over which nothing more than a horse had even been driven, changed from impossible to downright impassable, but we kept on, inspired by memories of the Buick. If in eleven years the company Buick could be taught to do everything but climb trees, we could teach this Ford truck a few tricks.

The trail ended at the Hat Ranch with a dry wash to cross just before we reached the corral. After what we had put the truck through to here, it mistook the dry wash for paved road.

The house, the scattered buildings, the corral itself, were beautifully in keeping with the trail to the place. It wasn't reasonable to expect to find a caretaker in charge, but there was one and we found him. He was able to tell us where to find the tracks we sought, so we pushed right on.

Following directions over a stretch of road no worse than we had been used to, we parked and walked up a short rise. On a windswept exposure of flat rock we found our tracks. They were indeed not dinosaur tracks; they were in fact almost a fresh trail by comparison. In the early days of the mammals, shortly after the end of Cretaceous time, the world was full of odd, flat-footed mammalian creatures of many types, before mammalian life had fully diversified. The carnivores learned to get up on their toes and spring. The herbivores learned to stand on very tiptoes and learned to run faster than carnivores. This was the age, the almost nowadays times, in which the tracks we found were made in fairly recent muds. There were tracks like those of modern deer, something that might have been made by a present-day puma, and a rather odd trail Barnum said was probably a ground sloth. There was an extensive and worth-while camel trail, and the tracks of a quite small bird, about the size of a robin. These were different from the tracks to which we were accustomed, and they definitely wouldn't be appropriate on the wall of Dinosaur Hall. Besides, the formation in which we found them made it impossible to lift samples for Brown's track collection. This was, after all, our vacation, and the task of lifting them wasn't suited to the time we had left. We borrowed a motto common among rock clubs and often found along nature trails: Leave nothing but footprints; take nothing but pictures. I did exactly that. November was coming at us, and Barnum was waiting in the truck.

37

I WATCHED THE LINE of heavy vans rumbling into the museum courtyard, van after van after van bringing in the mountain of crates and boxes Brown and I had packed in Glen Rose and had last seen as a tarpaulin-draped mountain in the lumber yard there. By truck and freight and steamer from Galveston, the long season's gleanings from the Paluxy and from the Big Bend country were home at last. Mid-December had found me, weary and out of breath, catching a train for New York, but it was only at this moment, two weeks later, that I felt the Texas job was done.

Came an ordinary van bearing an extraordinary big box, a box with a big 150 scrawled in black paint on its white plaster. Five of Steve Murphy's men leaped to the lowered tailgate, struggled with care to get the box onto the platform. It was the main box of those containing Barnum's big crocodile from the Big Bend. He had suggested getting to work on it right away. "We ought to see if we have anything worthwhile," he told me. "You can get on that job first, if you'd like."

I told the unloaders to set the big box where it might be readily moved to the laboratory. Other boxes and crates were unloaded and checked against the list, until we had all of the crocodile in its several parts accounted for, gathered together, and set aside. The tremendously bigger pile of boxes from the Paluxy could sit here in the court until we got to them; the crocodile had priority for now.

I was familiar only with the faraway hole in the hill down in the Big Bend where Barnum and Erich had dug it up; when they had dug and boxed it, I was knee-deep and better in the Paluxy. All I had go to on was the exterior appearance of numbered irregular blocks, with almost nothing in their shapes to relate them one to another. Three

sizable masses of bone in matrix, plastered and formless except as gobs of mud have form. About a dozen smaller jacketed pieces. An uncounted number of wrapped fragments. Altogether, seven or eight hundred pounds. I looked forward to starting the lab work in the morning. I doubted I would be all done by lunch-time . . . almost any lunch-time.

For the first time in a year and a half, I took my place again at the long table by the big window. Almost like home again. Certainly, there is more variety and adventure on the outside, but variety is the spice. And, besides, now it was winter.

Carl Sorensen, good old Carl, helped me unpack the two largest boxes. We rolled the heavy blocks out onto straw packing and piled the smaller lumps in trays. The very largest block we placed on a small revolving table for immediate investigation. Stripping away the plaster bandages, I found I had a shapeless mass of concretionary lime with almost no bone in sight except at one end. Another of the blocks had presumably made contact where the broken cross section showed. I collected cold chisels of assorted sizes and went to work. I had a hammer laid out for fine work, but the job looked as if it might require a heavier one I borrowed from Carl. I struck the mass lightly with the heavy hammer. Nothing happened; the thick limestone coating was as hard as flint. So I struck violently. The hammer clanged and bounced as if from an iron anvil. So I took a chisel and spalled a chip from a rounded corner. It hummed across the room like a ricocheting bullet.

Carl, standing by, said, "If that's the best you fellows could find in the Big Bend, I'd hate to see the worst."

I chipped off another flake, which skittered off under the nearby tables. The next done didn't stop until it clicked against a wall case beyond. Another hit the icebox. There was no way of controlling the flight of the chips, nor was there any other way of getting this job done. When I had chipped two inches of flinty limestone off the corner without sight of bone, I realized the size of the job ahead. It would take weeks to reduce these blocks and to work out the material hidden within.

I turned to examine the wrapped fragments, the odds and ends that Barnum had found lying on the ground. The first package held a meaningless lump of fossil bone. The next package held part of a huge tooth. It was heavy and conical, and in the eight-inch-long mass I recognized a typical crocodilian tooth, but for the size. I had never seen one more than half the size. Other packages contained other teeth, plus an unidentifiable piece of broken bone.

Brown came in while I was still marveling at the teeth, teeth larger than those of the tyrannosaur downstairs.

"One of these teeth made a direct contact with the block you have on the table," he said.

We tried the teeth, one by one, all around the block, but we couldn't find a fit anywhere. Brown moved off on other work, and I chipped and chiseled away the morning. Willie Booth came through with his broom, scowled at the litter on his clean tile floor, wiped a few chips off the top of his icebox, and went on.

The flinty stuff was a special limestone; it didn't flake off at the bone surface as good matrix should; instead, it seemed to have penetrated the bone as freezing water would penetrate a porous rag. Only by the feel of the chisel could I tell when I had reached bone. This, of course, did the bone little good. By late afternoon I had cleared off the upper side of as much bone as I could cover with one hand. Barnum wandered back in and grinned at my willingness to knock off and talk.

"What do you have now, R.T.?" he asked.

"It doesn't look like anything but a lump of bone."

He took out his magnifying glass and studied it over with care.

"Part of a skull," he pronounced. "Or maybe part of a jaw. Nothing to do but keep on keeping on."

Brown kept on, about other things. I kept on there. At four-thirty Willie Booth came back. His floor was well-littered with chips all over now.

"Except for Otto, this place is always clean until you come home," he grumbled. "Then it's always a mess."

At the end of a solid week's work, I had six irregular heavy lumps of bone to show for my troubles, plus enough chips to give Willie Booth overtime if he had known how to handle it. The pieces were only roughly cleaned, and they looked like part of a broken jaw. If I could have had the crocodile here alive, I'd gladly have broken his jaw for lying down to die in such a lousy limey spot, when there must have been a patch of clean sand around somewhere if he'd only bothered to look.

Barnum wandered by daily, viewed the creeping progress with patient interest, always moved on with the same cheering words, "Well, R.T., you'll just have to keep on trying."

I took the jackets off several other blocks, trying to find an easy one just to get a change of pace. A round lump, some eighteen inches across, partly flattened on one end, was my choice. A lime coating covered it, like the rest, but the deposit wasn't as thick. I decided to work it out on a sandbag, but it fell apart at the first touch of the chisel, broken along a hidden fracture. I was put out by the accident, but the break exposed cross sections, and allowed me to see what was bone and what was matrix. In places the grey lime even freed easily.

I cleaned the four freed pieces and cemented them together. When this hardened, I started to refine the surface; unbroken margins indicated a complete bone. While waiting for the pieces to set, I worked out a similar specimen. Both pieces resembled bones from a crocodile pelvis. Brown recognized the analogy the moment he saw it.

"R.T., you have an ilium, I do believe!" he said.

He left the room, and returned shortly with its presumed analog from a modern crocodile. The resemblance was not striking, but it was fairly close. A large depression on one side suggested the acetabulum or socket from which a hind leg hung. The bone was three times the size of that of a modern crocodile. The identification did not strike us, after expecting skull elements, but at least we had something to go on. I continued to chip away small fragments of matrix still adhering to the bone.

One portion of the bone was still grey with matrix. The hard lime, penetrating as with the other fragments, gave the mass a false shape. Monday morning I cut into a round depression on a wide edge of the bone. A small brown fleck of tooth enamel appeared. I could hardly believe my eyes. The chisel had penetrated what seemed to be the matrix-filled cavity of an old tooth socket! To be sure the occurrence was not a freak chance or an unnatural deformity, I cut deeper into the socket, and recovered a convincing sliver of enamel. I looked with new insight at the bone mass. It was not ilium; it was part of a skull. The revelation was astounding; the long-dead creature's old bones had turned on me . . . end for end.

About here, Barnum came in. I laid my tools aside and let the specimen speak for itself. I just helped a little.

"Our ilium has developed teeth," I remarked casually.

Barnum looked at me sharply; he thought I was joking. Then he looked at the bones.

"No, it can't be!"

"Oh, but it has. Look at that enamel." .

He got out his magnifying glass and examined the evidence a long moment.

"You're right, R.T. But what does that make this bone?"

"Could it be premaxillary?"

"Premaxillary!"

Brown stared dumbfounded at the fossil. The premaxillary is a rounded bone that united with an opposite to form the front portion of the reptilian jaw, which in the crocodile is the end of the snout.

"Wait a bit, R.T."

Brown hurried away, to return presently bearing the skull of a *Crocodylus porosus*, the largest of the modern crocodiles, known from the Malay Peninsula to the Philippines. It was a huge skull, almost three feet in length, with premaxillaries like a pair of teacups. The bone on the table was depressed on one side, rounded on the other. It was shaped like a premaxillary. The other similar bone, once presumed to be ilium, lay on a nearby table. We set the two pieces up side by side. Like setting two wash basins together. The end of the big croc's snout . . .

Brown looked at it, this way and then that way. Finally he said, "Well, R.T., just keep on digging."

He left the modern crocodile with me for reference. I went back to my task, glowing. Barnum had done it again. Whatever we had here, we had something special. For the first time in weeks, I didn't want to lay down my tools at quitting time, not even when Willie came in to sweep.

Next morning Brown brought in Charles M. Bogert, a foremost authority on living reptiles. Brown and I stood the two premaxillaries together as before.

Bogert looked at the *C. porosus* and back at our specimen. I expected him to catch his breath, which he did. Then he looked at me and laughed.

"Do you believe that?"

Others in the lab stopped by with a new interest. I began to develop a bone that was suggestive of a portion of the side of the creature's skull. Every square inch of cleaning seemed to take hours, but the day went quickly. Again I littered the floor with chips. Again Willie Booth grumbled by, but grinned when he saw the fearsome creature that was emerging from the chips. "Just the same, you're getting chips in the icebox!" he said.

Most of the pieces taken from the two larger blocks proved, when finally hewed from the stubborn matrix, to belong to the jaws. Two bones fitted together to form the point of the jaw. Another made up part of the dentary with its tooth sockets, but the rest were incomplete or difficult to place. I dismantled the jaw of *C. porosus*, and with the seven elements of one side lying before me, looked for new resemblances among the unplaced pieces. I was able to recognize sutures and margins in these pieces that either at once or bit by bit allowed me to identify their proper places in the jaw. They fitted roughly and poorly in places but furnished a guide to each bit's place in the jaw, as well as establishing the jaw's length.

The day came when there were no more pieces left to be worked out of the matrix. Our conviction that we had something special was muffled by the knowledge that it had no exhibition value in its present state; there was little for the layman to comprehend in the broken fragments. If we wished to mount and display these bones, they would require a lot of plaster filling and patching. I discussed this with Barnum.

"The idea's fine," he said. "But do you have enough pieces on which to base restoration? Can you carry it out?"

I didn't know for sure. Brown left me to my own devices. The fragments of the lower jaw, the only logical starting point, laid out together on my table, stretched out like a train of cars. I had, in fact, to butt another work table up against mine to make room. Modeling clay held pieces temporarily together and filled in missing pieces. It was a mess, but it was a more effective mess than I had hoped for. Brown came in just as I lined up the last pieces, and laughed.

"Keep on!" he said. "But no one will ever believe it."

Brown, Bird and Erich Schlaikjer with the reconstructed skull of *Deinosuchus* (*Phobosuchus*). Photo by Charles H. Coles. Courtesy Department of Library Services, American Museum of Natural History.

Bird (far right) at work in the preparation laboratory of the Department of Vertebrate Paleontology at the American Museum of Natural History. Courtesy Department of Library Services, American Museum of Natural History.

The length of the clearly established side of the jaw was six feet. The other side was quite incomplete, but the angle of union at the front showed the spread at the rear to be thirty inches, a nice-sized throat for rough and gluttonous eating. In the next few days I restored in clay both sides of the lower jaw and turned my attention to the skull. The huge premaxillaries were placed in position on the point of the lower jaw. On the top of the cranium there wasn't a single clearly identifiable piece, but there was a large mass of bone we had long ago recognized as part of the maxillary and with it, a portion of the palate. Together, the two gave a fine idea of the skull nearly as far back as the eyes.

Based on *C. porosus* plus what we had plus what could be guessed, I laid plans for rebuilding the skull. It must be constructed of clay on a framework of wire mesh and wood. I placed the maxillary in place, propped up on wooden blocks. More blocks were stacked in a great pile to support and shape the wire mesh. I spent the better part of a morning scouring the museum for clay. It took

pounds to spread an inch-thick layer over the mesh, but I ended with something quite satisfactory. Brown grinned broadly at what I was doing, offered several suggestions, brought in every type of crocodile skull in the museum's storage bins. The job of giving rebirth to this monster seemed endless, but as the days passed it grew more and more lifelike. I had lots of help and advice. Bogert was almost a daily visitor. Schlaikjer was around every Saturday morning. Otto Falkenbach, no mean hand on skulls of all kinds, was always available. Paleontologists E. H. Colbert and G. G. Simpson came through every day or so. Barnum was at my elbow so frequently it seemed he was always there. One morning he looked over the top of the cranium.

"R.T., I think you have it just a little high here. Try dropping it an inch or so."

I dropped it more than an inch and gathered opinions. Most favored lowering, but I moved it slightly higher just to assure myself I agreed; then dropped it back.

Harry Raven of Comparative Anatomy offered advice about the positioning of muscle at-

Bird and Brown at Big Bend, Texas, 1940. Courtesy the Texas Memorial Museum, University of Texas at Austin.

tachments. "I have a partly dissected crocodile skull I believe you've never seen," he said. "Would you like a look at it?" The number of pickled animals in Raven's collection was endless, but I didn't remember the crocodile. I followed him to his department, into a small store room I'd never seen, dark and filled with the stench of formaldehyde and other preservatives. When he turned on the light switch, I saw the floor was covered with earthenware jars.

"My reference library," Raven explained. He looked down the rows of labels. Presently he raised a lid and brought up a crocodile head, dripping and stinking of formaldehyde, for me to see. The hide on one side was stripped away, and all the muscles exposed. I forgot the smell as Raven pointed out the attachments. Pits and cavities on my model took on new meaning.

With the restoration nearly complete, new visitors were added to the old ones. There was Dr. Ed Weyer, editor of *Natural History Magazine*. Charles R. Knight, well-known artist and painter of prehistoric animals, stayed and talked more than an hour on several occasions. On his last visit, he stood off and eyed the skull speculatively.

"Are you going to mount it with the mouth open?" he asked.

I told him I hadn't decided. He stepped over to the paper-covered table, picked up a pencil, and sketched a hungry-looking crocodile on the brown paper. "You ought to make him look something like this," he suggested. That settled it.

On one of Raven's calls he told me of a great *C.*

porosus he had taken on the Malay Peninsula, using a dead monkey suspended on a rope for bait. The crocodile got both the monkey and the rope. Later Raven climbed the same tree with a rifle, waiting for the same fellow to appear.

"He rose to the surface just beneath me, and he seemed to be measuring me and the distance, wondering if he could make it up without a rope. When I got a bead on him and pulled the trigger, I was happy I didn't have a misfire," Raven said.

I wondered how this fellow on the table might have handled the situation. In life he would have been forty to fifty feet long. He might not have needed a rope.

Carl and Otto cast my clay pieces in plaster of Paris. Everything was put into place and the whole piece moved in beside Charlie Lang's forge and anvil for supporting iron work. We all decided finally the great fellow must meet the public with his mouth full open. He was a tremendous hit as a temporary exhibit in the foyer downstairs and continued to be popular when he took up permanent quarters in Cretaceous Dinosaur Hall. The carnivorous dinosaurs, of course, never mastered the water. But a few of these fellows along the shore must certainly have made some of the smaller plant eaters watch their steps.

Brown and Erich, who had dug the fellow up, were in full agreement on giving him a name: *Phobosuchus*, meaning fearsome crocodile. And Harry Sinclair had funded the Big Bend trip. What better name than *Phobosuchus sinclairi*?

Brown and Bird at Big Bend, Texas, 1940. Courtesy the Texas Memorial Museum, University of Texas at Austin.

38

IN MY ROOM in the Department of Geology I bent over a smoked paper record from the museum's big seismograph. Dinosaur hunting, like many other things, had been put aside for the duration of the war. I hadn't touched a fossil for months. Instead, I had taken over routine duties quite like a man in the army. I had no fault to find with my new assignment except that it was confining and too much like writing. Always, I would rather do it than write it up, whatever it was.

There had been a mild earthquake down in the Virgin Islands. The graph on the table showed a tremor quite similar to many others from the same region. I hadn't even felt it; the machine felt it, and I recorded it. Here it was May. Not the May I had known in other years, the May of bustle and plans and excitement, of dashing about and tying up loose ends and thinking up things we had forgotten and getting set for take-off on the next big long-planned or oft-delayed expedition. Barnum was on the Board of Economic Warfare in Washington. Others of my old department, Erich Schlaikjer, George Simpson, had taken commissions in the army and were off to parts unknown. Around the Explorers' Club, the story was much the same: there's a war on. And I was tending a seismograph and counting squiggles on its recording tape. And most days, it didn't even squiggle.

I had received a communication from Washington, stressing the need for "officers more or less professionally acquainted with deserts and desert conditions," but I didn't feel I was good army material. Any army I had to depend on was better off with me on the outside somewhere.

As I sat there one day, feeling a touch of the blues, a knock came at my door. I looked up from an uninterpreted P-wave on the tape to see Harold Vokes's welcome face. It had a fine grin on it, which made it even more welcome.

"R.T., how would you like to go away out West this summer, and prospect for vanadium?"

How would I like to go away out West? What did I know about vanadium? Sounded exciting. But it takes more than excitement to make a good prospector. Even when the alternative was reading teensy squiggles from a stupid seismograph.

"Have a chair, Harold," I told him. "Have two or three chairs. Take them with you, if you like. What do I answer first? And what the heck are you talking about?"

Harold nested himself comfortably in a chair, fully aware of the effect his words had on me. Then he went on.

"There's a mining corporation downtown, right here in the city, in urgent need of field geologists. They want anyone who can qualify to go out on this vanadium project. They called me up this morning and made a proposition. I had an idea you might be interested too."

I was still wondering what I knew about vanadium. I had seen samples of it, of course. I was aware it sometimes occurred in the Morrison Formation, along with dinosaur bones and petrified wood. Its occurrence was not common; beyond that, my knowledge was limited or nonexistent.

"Well, Harold, I hardly know what to say . . ."

Harold laughed, and gave me the old Vokes wink.

"Anyway, why not go down and say whatever to the head geologist? He would like to meet you, anyway. I told him you might be down."

He mentioned a telephone number in the business district and urged me to make an appointment before the week was out. I put the number down in my memorandum book, with certain reservations. I knew nothing of formal mining engineering, for all my digging holes in the ground after bones. But I did know the rocks in which vanadium was found. I knew weapons manufacturers were clamoring for steel. I knew that in certain steels vanadium was an absolute essential as a hardening agent. I knew it was a very rare natural resource, vital to the war effort. Assisting in the project as outlined by Harold, using my years of experience in geology, would be of far greater benefit than anything I could possibly accomplish in the army in company with eleven million other people. But could I meet a mining corporation's standards? I was so reluctant to hear a negative answer that I waited until the week was nearly out before making the appointment.

On the day of the appointment I walked into the reception room of a large suite of offices, high in a downtown skyscraper. The girl at the desk took my name in and returned almost immediately, smiling.

"Please go right in, Mr. Bird."

The man behind the desk looked as expansive as the desk behind which he sat. He rose to greet me, shook hands, and waved me to a chair. Turning to the girl, he said, "Send in Mr. Blanke."

Mr. Blanke appeared, moving with brisk efficiency. I sensed that here was the right-hand man in meeting potential employees and passing judgment on their professional abilities. Blanke shook hands and drew up a chair.

"I understand from Dr. Vokes you might conceivably be interested in our work. As you know, we are undertaking a complete survey of the Morrison in all areas where vanadium has been known."

He continued at some length, describing the localities. Then he paused, waiting for me to speak.

"Well, I've spent a number of years prospecting for bones in the Morrison and other Mesozoic formations, with Dr. Barnum Brown . . ." I rambled on, outlining some of my experience. The man behind the desk cleared his throat.

"It looks as if you're the sort we want."

Blanke seemed to be waiting for this cue. "What would you consider a fair salary, Mr. Bird?"

A fair salary? I hadn't even begun to think of a salary. Money never bothered me much; it came from the museum in bimonthly checks, and I turned it over to the bank on returning from trips and at other odd times, and my biggest money problem was misplaced or forgotten checks which caused the museum office from time to time to jog my elbow about clearing a check so that they could clear up their books. As for salary, I'd work for free and beg for food just to get out in the field again.

"Well-l-l," I said, "I've never been a hand at driving a bargain . . ."

Blanke misinterpreted my hesitation. He named a figure that to me seemed princely. A few minutes later I found myself walking out of the building, toward the subway, a field geologist for the mining company.

The next day Harold and I began to lay plans. We arranged for a six-month leave of absence from museum duties. I went to the museum library and read up on vanadium, made a quick trip out to Rye to say goodbye to Father and my sisters, and got out and dusted off my prospecting pick. I hoped to find dinosaurs in the Morrison, along with vanadium.

A day or two before we were to leave New York, Harold and I paid a last visit to the offices of the mining corporation to make final arrangements. Now came a startling surprise: we were required to sign papers pledging utmost secrecy concerning certain phases of the project. Vanadium, while it was indeed needed, did not compare in importance with another element, uranium, which we were to be hunting at the same time. The potential of this element was staggering. In prospecting for it, we would not merely perform a service to our country at war, we would assist in developing new sources of a power hitherto undreamed of, which, in the end, might turn the very tide of the war itself.

On a designated afternoon, Harold and I went down to the airport terminal opposite Grand Central Station. Mr. Blanke had made all reservations. Our tickets called for the highest wartime priorities accorded to civilians. The man who handed them to us across the counter chuckled. "You fellows must be somebodies! Nobody but a commanding general can bump you off the plane. Or you might have to give your seat to a lady if Mrs. Roosevelt came along."

Our bags were plastered with important-looking tags, and porters took them with swift reverence. We left LaGuardia aboard the five-thirty p.m. flight for Salt Lake City and settled ourselves in adjacent reclining chairs. How different this trip was from the first trip west ages and ages ago on my old motorcycle. I wondered how many other geologists were being picked up and whisked about on the same mad secret mission. I still had doubts about my usefulness as a mining engineer, but as a prospector of the Morrison . . . well, if I couldn't hold up with the best, then Barnum

Brown had wasted an awful lot of valuable time.

When morning came, I pushed aside the curtains. Outside, the silver wing of the ship, motionless and glittering in the morning light was suspended in an ocean of air that rushed by us. I leaned closer and looked down. At first, darkness. Then earthbound shadows resolved themselves into side-lighted rounded hills, and a small town. Rawlins, Wyoming! Out of all the possible spots over which we might have flown, here we were above this small but well-remembered landmark. Two thousand feet below, Point of Rocks was going by. The Mesaverde was there, grey and dark, exposed along its demarcations between the low escarpment that down below was the impressive Golden Wall and the thin blue shales of its upper beds. The Ericson sandstone sprawled out, lighter than the other rocks. I picked out the black thread of the highway following the cut made by Bitter Creek and the gleaming steel of the Union Pacific swinging in toward Thayer Junction and Horse Thief Canyon.

The air beacon winked from Point of Rocks, still bright in the semi-dawn. Below its tower I saw the little building that housed the power plant, and I thought of the many times Barnum and I had eaten lunch in the cool shade of those white walls. Not far away was the little quarry from which the armored dinosaur had been dug out, in the moving panorama a pinpoint prick on the dim carpet of the Mesaverde. In my imagination I could still see Brown in his old hat and shirtsleeves standing on the edge of the excavation, helping the boys lower the bones down the slope to the road in a washtub. I felt a wave of homesickness for the old days and for long adventures with Barnum Brown.

A few brief hours, we would be back in the Mesozoic formations, not to uncover secrets of the past but to seek an element to dramatically change the future.

In the brightening day, the plane droned on west, taking us into a new life. For the old life, this was the end.

Roland T. Bird

Bird and his motorcycle.

Epilogue

R.T. BIRD'S HEALTH failed him in mid-life. On a uranium-hunting trip out West for the government, his body gave out. For the balance of his life, he considered himself all but clinically dead. But he learned to live with his "old body." The medical profession could not identify the cause of his physical breakdown, beyond suggesting that it was a possible result of his childhood bouts with rheumatic fever.

With strict physical and dietary regimens, R.T. lived another two decades. In his last years, I worked with him on experiments at making a copy of the lower jaw of the tyrannosaur found by Barnum. R.T.'s working day consisted of two or three stretches of work, each twenty minutes long. The rest of the day was spent in rest. Yet he achieved his goal. The copy of the jaw now rests in the little museum in Glen Rose, Texas, five or six miles from the spot where R.T. uncovered the sauropod tracks in the Paluxy bed.

In the early part of R.T.'s enforced retirement,

Dr. E.H. Colbert, then director of the American Museum in New York, wrote to remind him that a hundred or more broken blocks of the trackway from the Paluxy still lay in the museum's yard. The boxes were rotting, Colbert said, the plaster crumbling, most of the labels missing, the whole a weathering mess. Could R.T. have a try at putting it all together for display behind Barnum's brontosaur?

From some depth, he found the strength to do the job. Originally scheduled for about six weeks, it took six months. The broken, eroding blocks were put back together with an artist's skill. Barnum's brontosaur now stalks along in the museum leaving behind tracks that look like his, even if dinosaur and tracks come from different sites.

The display is a tribute to R.T. Bird, who found the tracks; to Ned Colbert, who saved them from destruction; and, of course, to Barnum Brown.

Applying base for dinosaur tracks.

Paluxy tracks and brontosaur skeleton at the American Museum of Natural History.

Paluxy River tracks at the American Museum of Natural History.

Notes

THE PURPOSE of these notes is to direct the interested reader to sources of additional information about topics discussed by Bird, to clarify potentially confusing matters, and to indicate where new knowledge about fossils has led to different interpretations than those given by Brown or Bird. In no case should any criticism of their ideas be taken as indicative of disrespect for the men or their work. Our knowledge of prehistoric life is inherently incomplete; in the words of St. Paul, "We see through a glass darkly." If today's paleontologists see any less darkly than Bird and Brown did, it is only because we have more information about fossils available to us than they did—and this new knowledge rests on the foundations laid by earlier paleontologists like Brown and Bird.

In preparing these notes and my introduction to Bird's memoirs, and in locating photographs for the book, I have been aided by many people: Hazel (Mrs. Roland T.) Bird, Peggy (Mrs. Junius) Bird, Mrs. Alice Bird Erickson, Dr. Wann Langston, Jr., Dr. John S. McIntosh, Dr. John H. Ostrom, Dr. Edwin H. Colbert, Dr. Philip D. Gingerich, Dr. Daniel C. Fisher, Mr. Ron L. Richards, Dr. Diane E. Beynon, Ms. Charlotte Holton, Dr. David Dilcher, Dr. Frances Brown, Mr. Walter Sorenson, Mr. Gregory S. Paul, Dr. Walter P. Coombs, Jr., Dr. Donald Baird, Dr. Dale A. Russell, Dr. Peter Dodson, and Mr. Chris Andress. Mr. Elmer Denman made usable copies of numerous old, yellowed photographs. During my work, I was supported by grants from the American Philosophical Society and the Office of Sponsored Research, Indiana University-Purdue University at Fort Wayne.

Before proceeding to specific chapter notes, it may be helpful to list some sources of information about general topics:

General Information About Dinosaurs:

A. Charig, 1979. *A New Look at the Dinosaurs.* Mayflower Books, New York.

E. H. Colbert, 1983. *Dinosaurs: An Illustrated History.* Hammond, Inc., Maplewood, New Jersey.

A. J. Desmond, 1976. *The Hot-Blooded Dinosaurs.* Dial Press/James Wade, New York.

D. F. Glut, 1982. *The New Dinosaur Dictionary.* Citadel Press, Secaucus, New Jersey.

L. B. Halstead and J. Halstead, 1981. *Dinosaurs.* Blanford Press, Poole, Dorset, United Kingdom.

D. Lambert, 1983. *A Field Guide to Dinosaurs.* Avon Books, New York.

J. H. Ostrom, 1981. *Dinosaurs.* Carolina Biological Supply Company, Burlington, North Carolina.

D. A. Russell, 1977. *A Vanished World: The Dinosaurs of Western Canada.* National Museums of Canada, Ottawa.

W. Stout, 1981. *The Dinosaurs: A Fantastic View of a Lost Era.* Bantam Books, New York.

R. D. K. Thomas and E. C. Olson (editors), 1980. *A Cold Look at the Warm-Blooded Dinosaurs.* American Association for the Advancement of Science Selected Symposium 28. Westview Press, Boulder, Colorado.

History of Study of Dinosaurs:

E. H. Colbert, 1968. *Men and Dinosaurs: The Search in Field and Laboratory.* E. P. Dutton, New York. Reprinted by Dover Publications, New York, 1984, under the title: *The Great Dinosaur Hunters and Their Discoveries.*

G. E. Lewis, 1964. Memorial to Barnum Brown (1873–1963). *Geological Society of America Bulletin* 75(2): P19–P27.

Stratigraphy and Earth History:

R. H. Dott, Jr. and R. L. Batten, 1981. *Evolution of the Earth.* 3d ed. McGraw-Hill, New York.

Chapter 1

1. For a beautifully illustrated recent account of the geology and paleontology of the Petrified Forest, see volume 51, number 4, 1979, of *Plateau*, the magazine of the Museum of Northern Arizona (Flagstaff, Arizona); the entire issue is devoted to this topic.

Chapter 6

1. Barnum Brown named Bird's amphibian *Stanocephalosaurus birdi* (see 1933, A new genus of Stegocephalia from the Triassic of Arizona, American Museum *Novitates* 640, 1–4). In contrast to earlier (Paleozoic) amphibians, many of which had had terrestrial habits, *Stanocephalosaurus* and its relatives (technically known as stereospondyls) were specialized aquatic animals. Their bodies were rather flattened from top to bottom, their limbs rather weakly built, and their heads disproportionately large. Some of these animals were the largest amphibians that ever lived; a European form had a skull about four feet long. Although very successful in the early part of the Mesozoic, stereospondyls soon became extinct, to be survived by the amphibian lineages that ultimately gave rise to the modern frogs, toads, salamanders, and caecilians.

Chapter 8

1. Brown published a description of the Howe Quarry work in the June, 1935 issue of *Natural History* (Sinclair Dinosaur Expedition, 1934, *Natural History* 36: 2–15). About 4000 dinosaur bones, representing at least 20 individual animals, were taken from this site. The most abundant dinosaurs at the Howe Quarry were sauropods, and most of these were tentatively identified by Brown as *Barosaurus* and *Morosaurus* (the latter now known as *Camarasaurus*); there were also whiplash tails reminiscent of *Diplodocus* or *Apatosaurus* (the animal formerly known as *Brontosaurus*). The ornithopod *Camptosaurus* was also present. A large carnivore (probably *Allosaurus*) was represented by isolated teeth. *Stegosaurus*, present at many Morrison Formation sites, was absent at the Howe Quarry. A technical description of this collection has never been published, and as a result some of the identifications (especially of the sauropods) are open to question.

The preservation of bones at the Howe Quarry is most interesting. There is no indication that the bones suffered water transportation; indeed, their preservation in a fine-grained silt strongly suggests that the animals died where their bones were preserved, in a quiet, low-energy environment. The bone layer was thinnest at the outer edges of the deposit, while in the central portion of the layer the bones were stacked two and three deep. Equally interesting, twelve dinosaur legs and feet were preserved standing upright, with the bones articulated.

Brown believed that the sauropods were water-living animals who perished as their aquatic habitat dried up. A recent study of the paleoecology of the Morrison (P. Dodson, A. K. Behrensmeyer, R. T. Bakker, and J. S. McIntosh, 1980, Taphonomy and paleoecology of the dinosaur beds of the Jurassic Morrison Formation, *Paleobiology* 6: 208–232) paints a slightly different picture. Sediments of the Morrison Formation accumulated on a huge plain that consisted of a variety of terrestrial and aquatic habitats. The occurrence of sauropod remains is not restricted to aquatic environments; rather, these dinosaurs seem (like other large dinosaurs in the fauna) to have wandered widely across the entire range of Morrison habitats. This interpretation is consistent with more recent interpretations of sauropod anatomy.

The bulk of the Howe Quarry assemblage probably represents the mass mortality of a number of sauropods (perhaps a herd) that became mired and died in the soft, fine-grained sediments of the river overbank or levee environment. The geometry of the Howe Quarry bone deposit suggests the possibility that the dinosaurs died where they did because they were attracted to a wet area during a dry spell, perhaps a drying water hole that became a sticky death trap.

2. One of the many perversities of the fossil record is the fact that sauropod skeletons so often lack articulated or associated skulls, as noted by Bird for the Howe Quarry. Presumably skulls became separated from vertebral columns during decomposition of the brontosaurs' soft tissues. This annoying situation was responsible for *Apatosaurus* (*Brontosaurus*) being assigned the wrong skull type for many years (see J. S. McIntosh and D. S. Berman, 1975, Description of the palate and lower jaw of the sauropod dinosaur *Diplodocus* (Reptilia: Saurischia) with remarks on the nature of the skull of *Apatosaurus*, *Journal of Paleontology* 49: 187–199).

Chapter 9

1. In 1907 Barnum Brown described gastroliths ("gizzard" or "stomach" stones) associated with a hadrosaur skeleton: "In 1900, while collecting fossils in Weston County, Wyoming, . . . I found a *Claosaurus* skeleton imbedded in a hard concretionary sandstone. In chipping off the surplus stone three rounded well-worn pebbles were found near the fore legs, embedded in the same matrix. These specimens were preserved and the occurrence made note of at once, for similar stones had not been seen anywhere in the deposit. These pebbles are rounded and vary in size, the largest measuring nearly three inches across" (Gastroliths, *Science*, N.S. 25(636): 392). In 1941 (The last dinosaurs, *Natural History* 48: 290–295) he reported additional occurrences of dinosaur gastroliths: an ornithopod from Mongolia with 112 stones in its body cavity, a *Barosaurus* skeleton

with seven "highly polished" stones preserved among its vertebrae, and the Howe Quarry sauropod described by Bird, which had "sixty-four well-polished stones under the shoulder blade." (In Bird's account, the gastroliths were found "between the pelvis and the abdominal ribs . . . all lying next to the abdominal ribs"—a slightly different position than reported by Brown).

Other workers have also reported gastroliths associated with dinosaur skeletons. Werner Janensch (Magensteine bei Sauropoden der Tendaguru-Schichten, pp. 135–144 in *Wissenschaftliche Ergebnisse der Tendaguru-Expedition, 1909–1912*, E. Schweizerbart'sche Verlagsbuchhandlung (Erwin Nägele) G.M.B.H., 1929–1935) described presumed stomach stones found among the neck vertebrae of Jurassic sauropods at the Tendaguru site in Africa; Janensch suggested that the gastroliths might have been vomited from the stomach during the dinosaurs' death throes. Gastroliths have been found in the gut regions of skeletons of prosauropods of early Mesozoic age in Africa (G. Bond, 1955, A note on dinosaur remains from the Forest Sandstone (Upper Karoo), *Occasional Papers of the National Museum of Rhodesia* 2(20), 795–800; M. Raath, 1974, Fossil vertebrate studies in Rhodesia: further evidence of gastroliths in prosauropod dinosaurs, *Arnoldia* (National Museums and Monuments of Rhodesia) 7(5): 1–7).

Despite Brown's observations, the matter of dinosaur gastroliths remains as controversial today as it was in his time. In a review of the occurrence of gastroliths in fossil reptiles, D. G. Darby and R. W. Ojakangas (1980, Gastroliths from an Upper Cretaceous plesiosaur, *Journal of Paleontology* 54: 548–556) concluded that unquestionable occurrences of stomach stones *within* the body cavities of dinosaurs, as opposed to scattered throughout the enclosing rock, are rare. For example, in his 1941 *Natural History* article Brown wrote: "In some fields where dinosaur skeletons are numerous, as in the Lower Cretaceous beds of Montana, we find literally thousands of highly polished stones that probably were regurgitated by dinosaurs after the stones became rounded and therefore no longer useful as grinders. None of these highly polished stones were found in the body cavities of skeletons from the same beds. We did, however, find such stones while excavating one of the skeletons, and they show the same high polish as those found exposed in the surface layers." Brown's observations indicate at least that *some* herbivorous dinosaurs *occasionally* ingested gastroliths, but it is difficult to determine whether this practice was widespread among the different kinds of dinosaurs, or even if it was frequent in those types of dinosaurs that did ingest stones. Perhaps more damning, pebbles similar to those that in dinosaur-bearing rocks are identified as gastroliths have been reported in rock units that variously lack dinosaurs, are unfossiliferous or were deposited before or after the time of the dinosaurs (see references cited by Darby and Ojakangas).

There is somewhat better evidence that plesiosaurs (a group of Mesozoic marine reptiles) made use of gastroliths. Darby and Ojakangas summarize earlier records (including one by Brown) and report a specimen (*?Alzadasaurus*) from the Bearpaw Shale of Montana in which 197 rounded pebbles of quartzite, chert, and smaller numbers of other rock types were found in what would have been the gut region of the plesiosaur; equally important, no other such pebbles were found in the shale outside the region of the skeleton. D. A. Russell (1967, Cretaceous vertebrates from the Anderson River N.W.T., *Canadian Journal of Earth Sciences* 4:21–38) described a skeleton of a small plesiosaur (*?Cimoliasaurus*) from the Northwest Territories of Canada and reported that "approximately 3 lb. of gastroliths were collected from an oblong area 10 in. in diameter immediately in front of the pelvic girdle. The largest stone is 60 mm long in its greatest dimension, while the smallest are only 8 mm in diameter."

Just as controversial as the occurrence of gastroliths in fossil reptiles is the interpretation of how these gastroliths were used. Herbivorous birds consume stones and sand (collectively termed grit) that accumulate in the gizzard, there helping to grind food when the gizzard wall is moved by muscular activity. The coarser the plant material eaten, the larger the particles of grit the bird consumes; ostriches are reported to ingest up to two pounds of grit composed of pebbles as much as an inch in diameter. Two hundred stones, weighing a total of 5½ pounds, were found in the stomach region of the skeleton of a moa from New Zealand (for a review of grit in bird stomachs, see A. L. Thomson (ed.), 1964, *A New Dictionary of Birds*, McGraw-Hill, pp. 341–342).

Herbivorous dinosaurs are presumed by many workers to have ingested gastroliths for similar reasons (see, for example, R. T. Bakker, 1980, Dinosaur heresy—dinosaur renaissance: why we need endothermic archosaurs for a comprehensive theory of bioenergetic evolution, pp. 351–462 in R. D. K. Thomas and E. C. Olson (eds.), *A Cold Look at the Warm-Blooded Dinosaurs*, American Association for the Advancement of Science Selected Symposium 28; see pages 419–420 of Bakker's article). Sauropods have ridiculously tiny heads for their body sizes, leading many workers to wonder how these reptiles could have processed food rapidly enough to meet their metabolic needs, particularly if one argues that dinosaurs had rapid metabolic rates (see J. C. Weaver, 1983, The improbable endotherm: the energetics of the sauropod dinosaur *Brachiosaurus*, *Paleobiology* 9:173–182). Consequently stomach stones are argued to have provided in-gut "mastication" of an intensity beyond that of which sauropod teeth were capable—but again, there is as yet no conclusive evidence that these dinosaurs did in fact commonly ingest stones (although I would not be surprised if this had been a common habit).

The function of gastroliths in living crocodilians may also be food comminution, as in birds. Unpublished observations of a caiman made in the late 1960's at Yale University by A. W. Crompton, K. M. Hiiemae, and M. Gibbons (summarized by Darby and Ojakangas) have been used to support this interpretation. A dead mouse was injected with radio-opaque dye and fed to the crocodilian. "For the first 36 hours very little gastric activity was observed and the gastroliths were neatly arranged in rows in the folds in the lining of the stomach . . . (after this) all hell broke loose. The gastroliths were moving like pebbles in a cement mixer and the mouse became completely invisible with the dispersal of the Lipiodol (the dye) throughout the stomach."

Against this interpretation of gastrolith function is the observation that the stomachs of crocodiles killed in the wild contain "entire bones in all stages of digestion" as well as delicate structures like the opercula (structures that close off the opening of the shell) of snails (H. B. Cott, 1961, Scientific results of an inquiry into the ecology and economic status of the Nile crocodile (Crocodilus (sic) niloticus) in Uganda and Northern Rhodesia, Transactions of the Zoological Society of London 29(4): 211–357); one would expect such objects to be pulverized by the kind of "cement mixer" operation observed by Crompton, Hiiemae, and Gibbons.

In observations on alligators and caimans, D. C. Fisher (1981, Crocodilian scatology, microvertebrate concentrations, and enamel-less teeth, Paleobiology 7: 262–275) reported that these crocodilians regurgitate as digestive residues packed masses (hairballs) of the hair, claws, feathers, and cuticle (of insects) of the animals they eat, just as hawks and owls regurgitate such items in pellet form (although the processes of pellet formation differ in crocodilians and birds). Bones and teeth become decalcified during digestion. These hard structures retain their form, however, because the organic matrix of the bones and teeth remains, and one can find and identify these now rubbery structures in the crocodilians' feces. It seems unlikely that one would be able to find more or less intact claws in the hairballs, or decalcified bones and teeth in the feces, of crocodilians if gastroliths were important in grinding their food. At least most of Fisher's captive crocodilians did not have stomach stones (Fisher, personal communication). This might account for the survival of bones and teeth in his animals' droppings. However, Fisher also dissected wild-captured crocodilians that did have stomach stones and observed patterns of hard tissue degradation like those seen in his captive animals; similarly Cott's observations on the condition of bones in the guts of large Nile crocodiles (which do contain gastroliths) suggest that Fisher's findings can be extended to free-living crocodilians.

Cott favored the hydrostatic ballast interpretation of gastroliths. He reported that the percentage of Nile crocodiles with stomach stones increases during growth such that all animals 2.5 meters or more in length carry them. The gastrolith load of adult crocodiles averages about 1% of their body weight, but can get close to 3% in animals with more stones than usual. The gastroliths have a greater effect on the crocodiles' submerged weight; a 1% gastrolith load by weight in air is equivalent to 12.5% of the animal's weight in water (25% for a crocodile with a 2.5% load in air). A standard 1% stomach stone load would increase a crocodile's specific gravity from 1.08 to 1.09.

Cott suggestd that this increase in specific gravity would enable a crocodile better to rest on a river bottom in the face of a strong current, or give the reptile an edge in dragging large prey underwater. However, I doubt that a roughly 1% increase in specific gravity would make that big a difference in fighting currents or large animals. In fact, one would think that a ballast load heavy enough to have a significant effect on these two situations would make a crocodile so negatively buoyant as to have problems at other times; the reptile would continuously have to resist sinking by muscular effort.

It seems more likely that stomach stones function in maintaining a crocodile's stability in water. Cott reported that young crocodiles, which lack gastroliths, tend to be top-heavy and tail-heavy; gastroliths may serve to correct this problem. Darby and Ojakangas believe plesiosaurs to have used gastroliths for similar reasons.

Chapter 14

1. The "new" ceratopsian jaw bone was the intercoronoid (B. Brown and E. M. Schlaikjer, 1940, A new element of the ceratopsian jaw with additional notes on the mandible, American Museum Novitates 1092: 1–13), a small bone associated with the inner ventral edge of the dentary, the tooth-bearing bone of the reptilian lower jaw. Bird's observation that the jaws of reptiles are poorly constructed is rather extreme. Reptilian jaws are quite capable of getting the job done, and the loosely constructed jaws of snakes are actually advantageous in permitting these reptiles to swallow relatively large prey items.

2. Daisie Adelle Davis Sieglinger, known to her followers (one of whom would be R. T. Bird) simply as Adelle Davis, author of such books as Let's Get Well, Let's Have Healthy Children, Let's Eat Right to Keep Fit, Let's Cook it Right, Vitality through Planned Nutrition, and You Can Stay Well, was for many years America's premier popular nutritionist. She was also, in the eyes of the medical establishment, a food faddist and a quack. She believed that a "proper" diet could prevent or even cure a host of ailments, including allergies, kidney disease, glaucoma, muscular dystrophy, multiple sclerosis, hemorrhoids, and cancer, and that better nutrition was the answer to mental illness, divorce, drug addiction, alcoholism, and crime. She died in 1974, astonished at

her inability to conquer by dietary methods the cancer that killed her. For details of her career, see R. M. Deutsch, 1977, *The New Nuts among the Berries*, Bull Publishing Company, Palo Alto, California.

Chapter 15

1. The big phytosaur collected by Brown and Bird from Tanner's Crossing, near Cameron, Arizona, was identified as *Machaeroprosopus* and described in a monograph published by E. H. Colbert in 1947 (Studies of the phytosaurs *Machaeroprosopus* and *Rutiodon*, *Bulletin of the American Museum of Natural History* 88:53–96).

Chapter 16

1. The remains of Pleistocene (Ice Age) mammals are common in northern Indiana, particularly in bog deposits, the setting in which Bird's mammoth specimen occurred. Although not as common as mastodonts (*Mammut americanum*), animals that were lower-slung and more stockily built than true elephants, mammoths have been found at a number of Hoosier localities. Four North American species of the mammoth genus *Mammuthus* are presently recognized; most or all of the Indiana mammoths probably represent *M. jeffersonii*. For details, see M. W. Lyon, Jr., 1936, Mammals of Indiana, *American Midland Naturalist* 17: 1–373; M. W. Lyon, Jr., 1942, Additions to the "Mammals of Indiana," *American Midland Naturalist* 27: 790–791; J. A. Holman, 1975, *Michigan's Fossil Vertebrates*, Michigan State University Museum Educational Bulletin 2; B. Kurtén and E. Anderson, 1980, *Pleistocene Mammals of North America*, Columbia University Press; E. L. Lundelius, Jr., R. W. Graham, E. Anderson, J. Guilday, J. A. Holman, D. W. Steadman, and S. D. Webb, 1983, Terrestrial vertebrate faunas, pp. 311–353 in S. C. Porter, ed., *Late-Quaternary Environments of the United States: Volume 1, The Late Pleistocene*, University of Minnesota Press; R. L. Richards, 1984, The Pleistocene vertebrate collection of the Indiana State Museum, with a list of the extinct and extralocal Pleistocene vertebrates of Indiana, *Proceedings of the Indiana Academy of Sciences* 93: 483–504.

Chapter 19

1. Barnum Brown described the dinosaur tracks from the States Mine, as well as the results of the rest of his crew's 1937 work in the Mesaverde Formation, in the March, 1938 issue of *Natural History* (The mystery dinosaur, *Natural History* 41: 190–202, 235). The length of the "mystery dinosaur's" step has been the focus of some controversy. Modern students of fossil trackways designate the distance between two successive tracks of the same foot as a stride, and the distance between two successive tracks of the opposite feet (i.e., left to right or right to left) as a pace (sometimes step). Unfortunately, Brown (in common with many older workers) used the word "stride" to describe what today is called a pace.

The States Mine trackmaker was a big ornithopod, as evidenced by its yard-wide footprints. Brown argued that if *Tyrannosaurus*, which in the American Museum mount stands 18½ feet high, had a leg length that would have permitted it to take 9-foot steps, then the mystery dinosaur, with its 15-foot steps, might have been 35 feet tall. An alternative interpretation, that the Mesaverde trackmaker had been of more modest (although still large) size, and running rather than walking (thus taking longer steps than usual), doesn't seem to have occurred to him, although some later workers did make this interpretation.

Even so, there is a question as to whether the step length of 15 feet represents a pace (as interpreted by Brown) or a stride. Between the two tracks in the portion collected by Brown and Bird is an amalgamation of two dinosaur tracks that were slightly larger than those in the trackway sequence, and which Brown believed were made by different animals than the trail-maker. One of these amalgamated tracks points in the same direction as the two footprints in the trail, and quite possibly represents part of the same trail. If so, then Brown's 15-foot step is a stride rather than a pace, and there is no reason to infer either gigantic size or rapid locomotion for the trail-maker.

Further clouding the picture, however, is Brown's observation that there was a third footprint in the trackway, beyond the two that were collected. Although Brown doesn't explicitly say so, one assumes that this track was about 15 feet beyond the second, or Brown would have stated otherwise. Perhaps the mystery dinosaur was taking 15-foot paces after all. For a further discussion of this controversy, see M. G. Lockley, B. H. Young, and K. Carpenter, 1983, Hadrosaur locomotion and herding behavior; evidence from footprints in the Mesaverde Formation, Grand Mesa Coal Field, Colorado, *Mountain Geologist* 20: 5–14. Lockley et al. report other Mesaverde ornithopod trackways with apparent pace lengths comparable to that reported by Brown, but emphasize that pace lengths are quite variable from one trail to another. Adding to the difficulties of interpretation, the narrowness of coal mine tunnels, and their poor lighting, make it difficult to trace particular dinosaur trackways very far or very confidently (Lockley, personal communication). Whether the trails with long paces were made by moderate-sized runners or much larger walkers depends on the estimate of the leg length/track length ratio used; Lockley et al. argue that this ratio was higher than generally believed, and so most of the Mesaverde trails were made by slowly-moving ornithopods.

Chapter 21

1. Although the claim that Barnum Brown was the discoverer of the Folsom points has been repeated else-

where (see, for example, his obituary notice in *The New York Times* [February 6, 1963], the account of his life given in the 1969 *National Cyclopaedia of American Biography* 51: 483–484, and G. E. Lewis' memorial to Brown in the *Geological Society of America Bulletin* [see bibliography]), the story is more complicated than that. The initial American discoveries of projectile points associated with fossil mammals were made at Lone Wolf Creek (Colorado, [Mitchell County] Texas), Russell Springs, Kansas, Frederick, Oklahoma, and Folsom, New Mexico. These finds were reported in 1927 in back-to-back articles in *Natural History* (J. D. Figgins, The antiquity of man in America, *Natural History* 27(3): 229–239; H. J. Cook, New geological and palaeontological evidence bearing on the antiquity of mankind in America, *Natural History* 27(3): 240–247) written by staff members of the Colorado Museum of Natural History; Figgins was the museum's director. In September, 1927, while in Utah, Barnum Brown heard from Figgins that a new point had been found at the Folsom site. Brown's group went to New Mexico and Brown himself removed the remaining sediment around the artifact, which had been left *in situ* by Figgins' workers (see B. Brown, 1928, Recent finds relating to prehistoric man in America, *Bulletin of the New York Academy of Medicine*, Series 2, Vol. 4: 824–828; B. Brown, 1928, Artifacts discovered with bison remains in New Mexico, *Natural History* 28: 556; B. Brown, 1929, Folsom Culture and its age, *Geological Society of America Bulletin* 40: 128 [see also the discussion of Brown's abstract on pp. 128–129 of the last publication]). Brown's contribution to the Folsom story was thus not so much discovery as authentication (cf. Brown's obituary notice in *Time* magazine, February 15, 1963).

For recent discussions of Folsom Culture and its place in American prehistory, see G. C. Frison, 1978, *Prehistoric Hunters of the High Plains*, Academic Press; G. C. Frison and B. A. Bradley, 1980, *Folsom Tools and Technology at the Hanson Site, Wyoming*, University of New Mexico Press; G. C. Frison and D. Stanford, 1982, *The Agate Basin Site: A Record of the Paleoindian Occupation of the Northwestern High Plains*, Academic Press; J. D. Jennings (ed.), 1978, *Ancient Native Americans*, W. H. Freeman.

Chapter 22

1. Brown clearly believed that he was the first to use plaster jacketing in the collection of fossil vertebrates, a claim repeated in his 1938 "mystery dinosaur" article. Brown began using plaster jackets in 1897. However, in 1876 Cope and C. H. Sternberg, while collecting Cretaceous dinosaurs in Montana, covered bones with strips of cloth and burlap soaked in a rice paste, which upon hardening afforded a great deal of protection to the fossils. The following year one of Marsh's collectors, S. W. Williston (under whom Brown later studied paleontology at the University of Kansas) proposed covering

fossil bones with a protective layer of paper strips soaked in flour paste; the same year another Marsh worker, Arthur Lakes, wrote to his employer to say that he had been coating bones with plaster of paris. By 1880 Williston's and Lakes' ideas had been combined, and all of Marsh's collectors were covering fossils with jackets made of cloth or burlap dipped in plaster. For details, see E. H. Colbert, 1968, *Men and Dinosaurs: The Search in Field and Laboratory*, E. P. Dutton, 83–84.

Chapter 24

1. For a summary account and list of the Bridgerian fauna, see D. E. Savage and D. E. Russell, 1983, *Mammalian Paleofaunas of the World*, Addison-Wesley Publishing Company, Reading, Massachusetts, pp. 95–99. Contrary to Bird's comments, true oreodonts do not occur in the Bridgerian fauna; possibly he is referring to another group of primitive hoofed mammals, the dichobunids. However, dichobunids are much smaller than oreodonts and have rather different cheek teeth, so it seems unlikely that they could be confused (P. D. Gingerich, personal communication). Nor is *Eohippus* (now known as *Hyracotherium*) a Bridgerian form, although another primitive horse, *Orohippus*, is present, although rare. Just possibly Brown's party was not in the Bridgerian at all but rather collecting at the next lower stratigraphic level, the Wasatchian, which is exposed in the region around Rock Springs (Gingerich, personal communication). *Hyracotherium* is quite common in the Wasatchian, as is *Meniscotherium*, a moderate-sized herbivorous mammal that one might easily mistake for an oreodont (Gingerich, personal communication).

Chapter 27

1. Brown believed Bird's beautiful palm frond to have come from a *Sabal* palm and compared the fossil to foliage of the living *Washingtonia* palm (see "The mystery dinosaur," op. cit.). *Washingtonia* has spines on the petiole (leaf stalk), however, which Bird's specimen does not, so comparison with *Washingtonia* is questionable. The specimen certainly represents a *Sabal*-like form, however (David Dilcher, personal communication). On the same rock slab with the palm frond were plant fossils that Brown identified as figs, waxberry, and an *Araucaria*-like conifer.

Chapter 28

1. Brown's turkey-sized "pygmy dinosaur" was named *Microvenator celer* by John H. Ostrom (1969, *Stratigraphy and Paleontology of the Cloverly Formation (Lower Cretaceous) of the Bighorn Basin Area, Wyoming and Montana*, Bulletin 35, Peabody Museum of Natural History, Yale University). Brown believed this dinosaur to have had very large jaws for its size, because its bones were associated with teeth that were three to four times

as big as one would expect to see in a coelurosaur the size of *Microvenator*. Ostrom believes that these teeth belong to another, larger Cloverly theropod, which he named *Deinonychus* (1969, *Osteology of Deinonychus antirrhopus, an Unusual Theropod from the Lower Cretaceous of Montana*, Bulletin 30, Peabody Museum of Natural History, Yale University).

Because of *Microvenator*'s small size, it is possible that this dinosaur is a juvenile rather than an adult; supporting this interpretation is the fact that some of the sutures in the bones of the vertebral column have not closed. However, the texture and the well-formed shapes of the preserved bones led Ostrom to conclude that *Microvenator* is either an adult or close to it.

2. John MacLary published a brief description of his Purgatory River dinosaur footprint site (1938, Dinosaur trails of Purgatory, *Scientific American* 156: 72). He reported three-toed tracks of bipedal dinosaurs as well as the odd tracks examined by Bird. In 1982 Martin Lockley and his colleagues began new studies of the Purgatory River site; Lockley's group discovered about 1000 footprints in 60 trackways, including at least four sauropod trails that extend over a distance of more than 200 meters. For a preliminary description of this site, see M. Lockley, 1984, Dinosaur tracking, *The Science Teacher* (January, 1984): 18–24.

Chapter 29

1. Dinosaur tracks in the Glen Rose Formation were first reported in 1917 by Ellis W. Shuler (Dinosaur tracks in the Glen Rose Limestone near Glen Rose, Texas, *American Journal of Science* 44: 294–298). This initial report was followed by discoveries of dinosaur tracks in much of central Texas: W. E. Wrather, 1922, Dinosaur tracks in Hamilton County, Texas, *Journal of Geology* 30: 354–360; C. N. Gould, 1929, Comanchean reptiles from Kansas, Oklahoma, and Texas, *Bulletin of the Geological Society of America* 40: 457–462; E. W. Shuler, 1935, Dinosaur track mounted in the band stand at Glen Rose, Texas, *Field and Laboratory* 4: 9–13; E. W. Shuler, 1937, Dinosaur tracks at the fourth crossing of the Paluxy River near Glen Rose, Texas, *Field and Laboratory* 5: 33–36; C. C. Albritton, Jr., 1942, Dinosaur tracks near Comanche, Texas, *Field and Laboratory* 10: 160–181; D. S. Roberson, 1973, *Valley of the Giants*, Baylor Geological Society (Baylor University) Field Trip Guide; W. Langston, Jr., 1974, Nonmammalian Comanchean tetrapods, *Geoscience and Man* 8: 77–102; S. A. Skinner and C. Blome, 1975, *Dinosaur Track Discovery at Comanche Peak Steam Electric Station, Somervell County, Texas*, privately published, 16 pp.; W. Langston, Jr., 1979, Lower Cretaceous dinosaur tracks near Glen Rose, Texas, pp. 39–61 in B. F. Perkins and W. Langston, Jr., *Lower Cretaceous Shallow Marine Environments in the Glen Rose Formation: Dinosaur Tracks and Plants*, Field Trip Guide, American Association of Stratigraphic Palynologists,

12th Annual Meeting, Dallas (revised 1983); J. O. Farlow, 1981, Estimates of dinosaur speeds from a new trackway site in Texas, *Nature* 294: 747–748; R. H. Sams, 1982, Newly discovered dinosaur tracks, Comal County, Texas, *Bulletin of the South Texas Geological Society* 23: 19–23 (the author incorrectly refers these tracks to sauropods). R. T. Bird published four popular accounts of his work on Texas dinosaur footprints (1939, Thunder in his footsteps, *Natural History* 43: 254–261; 1941, A dinosaur walks into the museum, *Natural History* 47: 74–81; 1944, Did *Brontosaurus* ever walk on land? *Natural History* 53: 61–67; 1954, We captured a "live" brontosaur, *National Geographic* 105: 707–722.

The most commonly encountered tracks were made by three-toed, bipedal dinosaurs. Some of these tracks have long, narrow toes, sometimes with wicked-looking claw marks at their tips; these footprints were made by large carnivorous dinosaurs. While it is seldom possible to say exactly which kind of dinosaur made a particular track, a good candidate for the Texas carnosaurian trackmaker is *Acrocanthosaurus*, a relative of *Tyrannosaurus*, whose bones have been found in Oklahoma and Texas. Other three-toed tracks have shorter, thicker, blunter toes; while some of these may be poorly preserved carnosaur footprints, others may be the tracks of ornithopod dinosaurs, perhaps *Tenontosaurus* or even *Iguanodon*. The quality of preservation of bipedal dinosaur tracks in the Lower Cretaceous rocks of Texas is highly variable; some footprints are barely recognizable as such while others are quite good. The sauropod tracks were probably made by a modest-sized brontosaur named *Pleurocoelus*, skeletal remains of which are known from Wise and Blanco Counties in Texas.

The environment in which the dinosaurs left their tracks was a vast, shallow-water, lime-mud-bottom (carbonate) portion of the continental shelf of the ancestral Gulf of Mexico. Seaward of this region, in deeper water, lay a chain of reefs built largely by peculiar extinct clams; to the landward the marine, carbonate environment gave way to sands and clays deposited by rivers (and it is in the latter sedimentary environment where most of the few dinosaur bones known from this area are found).

At times the water covering the carbonate shelf was withdrawn, perhaps in the wake of tropical storms, and dinosaurs could cross the exposed mud flats. Sometimes the mud was exposed long enough that the sun baked it, hardening the substrate and creating extensive systems of mud cracks. In most cases a track surface was exposed for only a few hours or days, and the footprints were buried by fine-grained sediments when the sea returned to cover the surface. After the sediments turned to rock, the contrast in sediment type between the track layers and the muds that filled in the tracks made it possible for present-day erosion to scour away the latter, exposing the footprints.

There are potentially thousands of dinosaur footprint sites in the Glen Rose Formation (and less frequently in other Lower Cretaceous rock units) in Texas and in correlative carbonate units around the edge of the Gulf of Mexico. Recently a spectacular discovery of a vast number of sauropod footprints was made in a gypsum strip mine near Nashville, Arkansas, in rocks of the DeQueen Formation.

2. The "man tracks" represent an interesting footnote to Bird's work on the dinosaur footprints of the Glen Rose Formation. The chief "man track" enthusiasts are "Scientific Creationists," fundamentalist Christians who believe that the Genesis account of creation should be interpreted literally. Since the geologic evidence for an ancient earth and organic evolution is clearly incompatible with such an exegesis, the Creationists attempt to prove that the earth is not more than several thousand years old and that the stratigraphic evidence of evolutionary change is an illusion. Proof of the coexistence of humans and dinosaurs would be invaluable to their cause, and so the Creationists have vigorously searched for incontrovertible human tracks in the Glen Rose Formation.

In a letter to Wann Langston, Jr. of the University of Texas, dated October 29, 1976, Bird gave his own perspective on this extraordinary story: "Do you perchance know of some of the allegations that have been heaped upon me in the past in regard to this matter? I almost blanch even now at the memory!

"There are, as you probably know, three kinds of tracks. There are the easily recognized product of hammer and chisel; then there are the kind that make some members of the present day human race ready to call you a liar and a cheat if you so much as attempt to tell them that no man lived in the Age of Reptiles. These are nothing more than simple depressions or eroded flaws in a rock surface with an accidental (and very slight) resemblance of human footprints . . . in the minds of these beholders.

"Finally there are those that might be called the real article, for they are tracks of a sort . . .

"Have you ever seen a hen or a rooster attempting to navigate in the spring of the year a yard more than ankle deep in freshly thawed mud? Let us consider a large predatory carnosaur wading in deep, almost fluid mud that also rises well above this animal's ankle. The feet sink deeply; as each foot is withdrawn with all three toes compressed together for easy extraction, muds that were extruded outward and upward on penetration, slump partially back, leaving only this character-less, oblong cavity.

"How did I ever get caught up in these persistent arguments (based on these and other "tracks") that hold that men and dinosaurs existed on earth at the same time together?

"I made the unfortunate mistake of describing all too well both the fake and the real variety in the first article I turned out for *Natural History* relating the discovery of the Glen Rose sauropod footprints. But these were the days when both Barnum Brown and I were dependent on the Sinclair Oil Refining Company for field work grants. As the company sometimes featured Barnum's planned expeditions in their national advertising, it behooved him later to write articles that would please old Harry Sinclair and soften him up for yet another season. That year Brown had been unable to get away from New York until late fall, and it had been I, in carrying out most of the season's plans, who had traveled for some four months on some of this bounty.

"Well, Barnum read the original manuscript, chuckled over the title, and gave it his blessing. When Editor Buddy Weyer passed on it, he did so with this well remembered remark, 'R.T., I wish more of your colleagues upstairs could write like this.' 'Thunder in his Footsteps,' intended for Sinclair's eyes as much as anybody's, went to press, with me trying to add a touch of mystery by bringing in the 'man tracks.' Who could dream of what might happen, what *did* happen!

"The first rumble of approaching trouble reached me in a little religious magazine mailed from a friend in California, *Signs of the Times*. It obviously represented the beliefs of those who accept the story of Creation in Genesis at face value. But the feature article, by one Clifford L. Burdick, was something of a shocker. He had visited my old Glen Rose quarry site, and while reluctantly crediting me for saying, 'No man lived back in the Age of Reptiles,' he sought to prove by distorted statements I could not deny the 'man tracks' were NOT human.

"Well, I didn't like this, naturally, but the small magazine had a limited circulation, and while I didn't hold with its beliefs, if it helped some in their worship of the Almighty, I could stand it. That R. T. Bird (and a 'Dr. Bird' no less!) hadn't been able to get around 'human' tracks in the early Cretaceous, was to laugh. I only hoped others who knew me would also laugh . . .

"Having started this story, I shall have to go on. Perhaps you already know how parties from these religious organizations came to dig for more man tracks at my old quarry site. They wanted proof of their own that such were here, and did not have much trouble establishing that truth when they struck that trackway of a carnosaur extracting his feet from deep mud with all three toes compressed together for easy withdrawal. The original muds at the area vary widely, from a hard sandy based type that retained poor prints, through those of perfect conditions to this fluid stuff where this one known carnosaur waded.

"Following these expeditions there in time emerged a respectable appearing book entitled *Man's Origins; Man's Destiny*. Eventually a copy came to hand from a friend in Pennsylvania. Much of the book's religious theme was tied to the man tracks, and again I suffered under the old aforementioned allegations. I was by now losing all patience with this 'Dr. Bird' thing, and while the author still gave me credit for stating 'No man lived back in the Age of Reptiles,' he did so reluctantly, going to great lengths to prove I didn't know what I was talking about, and condemning me for never again mentioning the man tracks in subsequent articles. The book was profusely illustrated by photos of sauropod and carnosaur tracks, together with those of the hand carved fakes and the mud wading carnosaur—both of the latter in his view true footprints of giant men.

"Would a day come when an author might drop my one defensive remark altogether? I couldn't believe it, but when the horrifying news came that a book now enjoying national circulation had done just that in the opening prologue, I almost fainted. Who was the author *this* time?

"With haunting memories of how I once mentioned a mysterious track without remembering seeing chickens slowly wading in deep fluid mud, pausing step by step to extract in turn each foot by folding the toes together for easy withdrawal, I entered our local Homestead library in fear and trembling. Was the book I sought 'in' at the moment?

"It was. It stood prominently displayed on a shelf of new books where it could not be missed by an incoming patron. All doubts about the author vanished now. Who, indeed, had not heard of Jeanne Dixon, the current outstanding lady seer, often consulted in greatly publicized circumstances to prophesy the futures of prominent Americans, the outcome of future elections. What had this woman wanted of *me*?

"Can you imagine my mental anguish as I read, 'Dr. Roland T. Bird of the Department of Vertebrate Paleontology of The American Museum of Natural History reported finding giant petrified human footprints along with those of dinosaurs in a river bed in Texas'? It was appalling to think of how many thousands of homes in America this book was invading at this very moment, from libraries and book stores everywhere.

"When I was fit to speak again, I fired off a scathing letter to this female 'seer' demanding an immediate withdrawal of my name from all future reprintings. In turn I received an immediate apology from her ghost writer . . . saying my name would never appear again in future editions . . .

"Time soothes the most painful of injuries, and again in this case, I was mollified by the usual religious overtones of the book. If her story should help bring some of her devoted followers back onto the paths of righteousness and the good will all of us should practice toward our fellows in this world here below, I would overlook the insult.

"As for those persons in vertebrate paleontology who might happen to see this book, I could hear them chuckling and saying to each other, 'Well, well, well, old R. T. Bird seems to have got himself into a jam here. The woman has, of course, grossly misquoted him. There must have been some flaw in his original story.

How did that escape old Barnum Brown? He and Bird always worked closely together.'"

The bulk of the Creationist effort has concentrated on a stretch of the Paluxy River upstream of Bird's quarry site, on land owned by one Emmet McFall. The most detailed account of their work is J. D. Morris, 1980, *Tracking Those Incredible Dinosaurs . . . and the People Who Knew Them* (Second Printing, Creation Life Publishers, San Diego). The Creationists have also released a widely circulated film about their work, *Footprints in Stone*.

For a long time the scientific community regarded the Creationists as nuisances that if ignored might go away. They didn't, and recently there have been several popular and semi-technical publications put out in response to Creationist arguments. With regard to the "man tracks," see D. H. Milne and S. D. Schafersman, 1983, Dinosaur tracks, erosion marks and midnight chisel work (but no human footprints) in the Cretaceous limestone of the Paluxy River bed, Texas, *Journal of Geological Education* 31: 111–123, as well as Langston's 1979 article on the Glen Rose dinosaur tracks (cited in previous note). The American Anthropological Association has released a videotape, *The Case of the Texas Footprints*, to counter the impact of *Footprints in Stone*. These works present interpretations of the "man tracks" similar to Bird's, with an interesting addition.

When an animal walks across a muddy surface, it not only leaves footprints in the mud layer it walks upon, but also distorts successive sediment layers below the actual track surface. Thus a single footstep produces a surface footprint as well as a stacked series of underlying "ghost tracks," each less distinct than the one above it—a phenomenon described by Edward Hitchcock as long ago as 1858. Deeper ghost tracks can have just the featureless, oblong shape so often seen in the Paluxy River "man tracks."

Thus it appears that those "man tracks" that are neither carved fakes nor irregularities in rock surfaces represent a class of poorly preserved bipedal dinosaur tracks. Some of these are probably collapsed tracks as described by Bird, and some are probably ghost tracks.

Chapter 32

1. Although Bird did interpret the Davenport Ranch sauropod trackways as having been made by a herd, it did not occur to him at the time that the herd might have been structured. Credit for this interpretation goes to R. T. Bakker (1968, The superiority of dinosaurs, *Discovery* [Peabody Museum of Natural History, Yale University] 3(1): 11–22), as Bird himself acknowledged in a letter to Wann Langston, Jr., of the University of Texas, dated August 10, 1976: "Bob Bakker did a spirited exercise . . . in which the sauropods come in for their share of comment. He mentions the Bandera-Davenport as-

semblage of sauropods, and ascribes to them the social behavior of animals arranged in a true 'herd' where the youngest and most vulnerable juveniles were sheltered in the center of the herd—a possible point I had never thought of." There is a growing body of evidence that suggests that the social behavior of dinosaurs was more complex than believed in Bird's day (see J. A. Hopson, 1977, Relative brain size and behavior in archosaurian reptiles, *Annual Review of Ecology and Systematics* 8: 429–448; J. O. Farlow, 1983, Dragons and dinosaurs, *Paleobiology* 9: 207–210).

Chapter 33

1. Although some early workers (most notably E. S. Riggs) believed the sauropods to have been terrestrial animals, the majority view was that these dinosaurs were aquatic creatures that spent most or all of their lives in the water. This is certainly the viewpoint expressed by Barnum Brown and R. T. Bird in the latter's memoir. It was believed that brontosaurs were too heavy to support their bulk on land. Furthermore, the elevated position of the nostrils on the sauropod skull was interpreted as providing these reptiles with a snorkel for breathing while submerged. Sauropod vertebrae are very lightly constructed for their size, with deeply excavated surfaces, while brontosaur limbs are massive. Some workers suggested that this distribution of skeletal mass—lighter in the dorsal region and heavier ventrally—would have enabled sauropods to wade through deep water without tipping over.

Although sauropods did enter the water on occasion (the Mayan Ranch trackway proves this), their skeletal anatomy makes equally good or better sense if these dinosaurs are interpreted as primarily terrestrial rather than aquatic animals (see R. T. Bakker, 1971, Ecology of the brontosaurs, *Nature* 229: 172–174; W. P. Coombs, Jr., 1975, Sauropod habits and habitats, *Palaeogeography, Palaeoclimatology, Palaeoecology* 17: 1–33). Calculations based on the cross-sectional area of sauropod limb bones suggest that these were easily strong enough to support the dinosaur's weight on land, and the sauropod trackways at Glen Rose and the Davenport Ranch corroborate this interpretation.

The position of the nostrils on the sauropod skull is of dubious meaning; while certain undoubtedly aquatic or marine mammals and reptiles have similarly placed nasal openings, others do not, and some living and fossil mammals known or believed to have had terrestrial habits have dorsally placed nostrils. Such a position of the nostrils is associated with the presence of a fleshy proboscis or trunk in some living mammals, and Coombs cautiously speculates that some sauropods may have had a proboscis of some kind. Equally troubling for the snorkel interpretation of dorsally situated sauropod nostrils is the fact that a submerged sauropod's lungs would be so far beneath the water's surface that it prob-

ably would have been impossible for the animal to inflate them against the pressure of the surrounding water.

The weight distribution of the sauropod skeleton does make sense as enabling a floating or wading sauropod to keep its balance. However, one might expect titanic land-living creatures to evolve in such a way as to reduce their weight to a minimum; the highly excavated sauropod vertebrae thus make sense as a weight-saving mechanism. On the other hand, one would expect the limbs of such huge animals, the structures that would have to support and move all that weight, to be massive. Thus the weight distribution within the sauropod skeleton does not prove that these dinosaurs were aquatic.

More to the point, there are features of the brontosaur skeleton that are contrary to what one would expect in aquatic animals. Many (but not all) aquatic and marine reptiles and mammals (such as hippos) have a body that is roughly circular or at least broad in cross-section, while large terrestrial animals (such as elephants) tend to have a deep (from top to bottom), narrow (from side to side), slab-sided cross-section. This latter condition permits more effective handling of the stresses associated with supporting a heavy body on land. Sauropods are clearly more like terrestrial than aquatic animals in their trunk cross-sections.

Aquatic and amphibious reptiles and mammals tend to have relatively short limbs (as in hippos); reduced or lost hind limbs (as in whales); elongated finger bones, and sometimes an increase in the number of fingers, to produce a flipper; reduction of all limb bones except those of the fingers into a compact structure, again associated with the construction of a flipper. All of these trends are contrary to the conditions seen in sauropods.

As a group, then, sauropods are more likely to have been terrestrial than aquatic animals; at the very least they probably spent significant amounts of time on land. This conclusion is consistent with the sedimentary environments in which sauropod fossils occur, as discussed in an earlier note.

Given the diversity of sauropods, however, it is probably unsafe to make hard and fast generalizations about the habits of all the brontosaur species. Some forms were probably more aquatic in habits than others. One must keep in mind, too, that the dichotomy between terrestrial and aquatic habits is somewhat artificial; "terrestrial" elephants frequently spend large amounts of time in and around water, while "aquatic" hippos make extensive nocturnal feeding excursions on land. Sauropods probably were comfortable in a variety of habitats.

2. Contrary to Bird's statements, the flesh-eating dinosaurs may have been reasonably good swimmers—perhaps even more comfortable in the water than sauropods. Emus, rheas, and cassowaries are capable swimmers, propelling themselves in the water by kicks of their three-toed feet (see E. T. Gilliard, 1960, *Living*

Birds of the World, Doubleday, New York, and articles on emus, rheas, and cassowaries in A. L. Thomson, 1964, *A New Dictionary of Birds*, McGraw-Hill, New York). Of the rhea (or ostrich, as he called it), the English naturalist Charles Darwin gave the following account: "It is not generally known that ostriches readily take to the water. Mr. King informs me that at the Bay of San Blas, and at Port Valdes in Patagonia, he saw these birds swimming several times from island to island. They ran into the water both when driven down to a point, and likewise of their own account when not frightened; the distance crossed was about two hundred yards. When swimming, very little of their bodies appear above water; their necks are extended a little forward, and their progress is slow. On two occasions I saw some ostriches swimming across the Santa Cruz river, where its course was about four hundred yards wide, and the stream rapid" (*The Voyage of the Beagle*, 1860 edition; Natural History Library, Doubleday, New York, 1962, 91). If flightless birds can swim in this manner, there seems no reason to suppose that carnivorous dinosaurs, with a rather similar foot and leg construction, could not (R. T. Bakker, 1971, Ecology of the brontosaurs, *Nature* 229: 172–174), and W. P. Coombs, Jr. (1980, Swimming ability of carnivorous dinosaurs, *Science* 207: 1198–1200) has described trackways from the Connecticut Valley that appear to have been made by swimming theropods.

Chapter 37

1. Barnum Brown published a brief account of the huge Texas crocodilian in 1942 (The largest known crocodile, *Natural History* 49: 260–261); a technical description was later published by E. H. Colbert and R. T. Bird (1954, A gigantic crocodile from the Upper Cretaceous beds of Texas, American Museum *Novitates* 1688: 1–22), who named the reptile *Phobosuchus riograndensis*. (The animal has subsequently been reassigned to *Deinosuchus*, perhaps *Deinosuchus rugosus*). The American Museum specimen consists of fragments of the skull and jaws; Bird reconstructed the entire head skeleton by assuming that the various bones had proportions similar to those in *Crocodylus*, as described in his memoir.

More complete material of a large crocodilian (but smaller than the American Museum specimen) from the same region was subsequently found, and is now in the collection of the Texas Memorial Museum. The more complete specimen has a more slender and possibly flatter skull, and a less massive lower jaw, than the Bird reconstruction. The teeth were closely spaced in the jaw, a spacing more like that seen in alligators than in crocodiles in the strict sense.

There is, however, some controversy over whether this second crocodilian is the same kind of animal as the one described by Colbert and Bird. In addition to the Texas material, remains of very large crocodilians have been found in Cretaceous rocks of Wyoming, Montana,

North Carolina, Delaware, and New Jersey (D. Baird and J. R. Horner, 1979, Cretaceous dinosaurs of North America, *Brimleyana* 2: 1–28). According to Baird (personal communication), the North Carolina material suggests a crocodilian with a relatively short and deep snout, more like Bird's reconstruction than the Texas Memorial Museum specimen. Baird believes that the American Museum and the Texas Memorial Museum crocodilians may belong to different species, or even genera.

Wann Langston, Jr. (University of Texas) disagrees (personal communication). Citing similarities in the premaxilla bone of the skull and the spacing of tooth sockets in the jaw between the two specimens, Langston believes that both of the animals are *Deinosuchus*, and that Bird's reconstruction is flawed.

Whatever the outcome of this dispute, it is clear that Late Cretaceous wetlands were haunted by at least one kind of gigantic crocodilian.

Large individuals of modern crocodilians are able to kill fairly large prey animals, such as deer, pigs, cattle, African buffalo, and camels (see S. A. Minton, Jr. and M. R. Minton, 1973, *Giant Reptiles*, Charles Scribner's Sons, New York), and there is an astonishing account (F. C. Selous, 1908, *African Nature Notes and Reminiscences*, Macmillan and Company, Ltd., London) of a Nile crocodile dragging a black rhinoceros into the water where it was drowned and presumably eaten. *Deinosuchus* probably reached weights comparable to those of many contemporary dinosaurs (a couple thousand kilograms or so). These observations suggest that *Deinosuchus* might have been a threat to even fairly large dinosaurs. Baird and Horner suggest that *Deinosuchus* may have been an even more important predator of hadrosaurs than tyrannosaurs were. **Please note late addendum:** The Baird-Langston dispute over whether the big Texas Memorial Museum crocodilian is the same kind of animal as the American Museum's *Deinosuchus* (*Phobosuchus*) has been resolved; Baird (personal communication) now agrees with Langston that it is. To quote Baird: "It's a croc with a long, low, alligatory snout that suddenly pinches in at the fourth-tooth notch and rises like the Rock of Gibraltar to the premaxillae. An odd thing entirely, but you can't argue with the specimen."

Chapter 38

1. During World War II, while prospecting for uranium, Bird began to suffer from heart palpitations—a condition that would trouble him the rest of his life. In addition, he started to be bothered by what he himself diagnosed as hypoglycemia. Whether he actually had the illness is open to question; Bird was a hypochondriac and a real eccentric when it came to food. He frequently put himself on diets of a single food item—potatoes, chicken, and fish patties all had their turn—that he would eat, meal after meal, for weeks at a time. It is

likely that many of Bird's health problems were due simply to poor nutrition.

After the war Bird returned to New York. He had decided that it was time to marry, and he began looking for a wife. As methodical as ever, he had a clear image of the kind of woman he hoped to marry, someone who would combine the qualities of Rowena Paton (by whom Bird had been more than a little smitten) and his brother Junius' wife, Peggy.

He found what he sought. Hazel Russell, introduced by a mutual friend, was the mother of two small daughters, Faye and Terry, adopted during a marriage that had failed. She was not like either Peggy or Rowena, but he decided she would suit him very well, while for her part she was captivated by this gentle, charming, intensely interesting man with his sapphire blue eyes and courtly manner. They were married in the Collegiate Church of St. Nicholas—he insisted that the ceremony take place in a Fifth Avenue church—on June 4, 1946.

R.T. brought his old motorcycle-camper out of storage, and late that fall he and Hazel took a camping trip to Florida. Thinking his health too poor to handle northern winters, Bird decided to move to Florida permanently, and he and Hazel built a home in Homestead, Florida. It was rather unusually landscaped, with replicas of dinosaur tracks from Texas put into the grounds. Returning to an enthusiasm of his youth when he had worked with prize cattle, Bird kept a cow and her calf on his property. Bird got along well with animals: the cow and calf played hide-and-seek with him and the girls, and he trained wild foxes to eat outside the door and whine for more.

In 1952 the administration of the American Museum decided that it was time to redo the brontosaur hall. The task fell to Barnum Brown's successor, Edwin H. Colbert. The skeletons of *Apatosaurus* (*Brontosaurus*), *Allosaurus*, and *Stegosaurus* were grouped together in the middle of the hall, specimen labels were rewritten, and a series of chalk sketches of dinosaurs was drawn on the walls of the hall above display cases. The most ambitious part of the renovation was the installation of Bird's Paluxy River sauropod footprints behind the brontosaur skeleton, something Brown had never gotten around to doing. Colbert asked Bird to return to New York for this project, and R.T. was more than willing.

The brontosaur's tail was removed and reinstalled, bent to one side to provide room for the tracks. Assisted by Carl Sorensen and two stone masons, Bird reassembled more than 100 pieces of Paluxy River bedrock into the trackways of the Texas sauropod and its carnivorous assailant. The work was not completed without incident. Bird's stone mason assistants were fond of taking a cola break; they would buy three soft drinks,

give one to R.T., and (unknown to him) lace the other two with something more potent. On one occasion, Bird, who didn't drink, got the wrong cola, and it made him rather ill.

R.T. missed Hazel terribly while he was away from her in New York. He wrote to her daily and was keenly disappointed if he didn't get a letter from her every day. Sundays made him very depressed, since there was no mail delivery.

Bird was a sentimental man. At one point his sister Alice, who was living in the family home in Rye, put up new wall paper in the entrance hall. This upset R.T. greatly; he asked his other sister, Doris, "Why did Sister (the name he always called Alice) obliterate the marks of our youth?" To Alice herself he simply complained, "Sister, this looks like dinosaur hide!"—in reference to the wallpaper's rough texture.

Although not a regular church-goer, Bird was a religious man. He was reared an Episcopalian and had been a choirboy, but as an adult he refused to take the Eucharist. He intensely disliked Franklin D. Roosevelt, whom he considered a hypocrite, and because of his unforgiving attitude believed himself unworthy of the Lord's Table. Because of his bad experiences with religious fringe groups, Bird became increasingly distrustful of the clergy. Hazel was a Presbyterian, and once her pastor made favorable mention of R.T. in a sermon; this both surprised and pleased him.

His health continued to be poor, for whatever reason. Adelle Davis, the well-known nutritionist, who had accompanied Brown and Bird on the Arizona field excursion during which the phytosaur skeleton was found, visited the Birds during a Florida lecture tour and was appalled by R.T.'s condition. She prescribed a diet for him, and whatever its real merits it was substantially better than his own self-imposed dietary regimen, and he perked up considerably for several years.

In the late 1970's he took another turn for the worse and was confined to his bed. During this time Hazel and Alice read to him from a book about dinosaurs by Adrian Desmond (see bibliography). Desmond's account of Bird's work pleased R.T.; he would remark, "I am listed among the titans!" Bird had developed a bad heart valve, and near the end it became infected. His doctors gave him massive doses of antibiotics in an effort to stop the infection, but to no avail. He died on January 24, 1978.

Roland Thaxter Bird is buried in Grahamsville, New York, a "gastrolith" in his pocket, surrounded by the Catskill Mountains that he loved as a youth. At the top of his headstone is carved a picture of a brontosaur; his epitaph reads, "Discoverer of Sauropod Dinosaur Footprints."

Index

Typesetting by *G&S, Austin*
Printing and binding by *Edwards Brothers, Ann Arbor*
Design by *Whitehead & Whitehead, Austin*